History of Physics

The Springer book series *History of Physics* publishes scholarly yet widely accessible books on all aspects of the history of physics. These cover the history and evolution of ideas and techniques, pioneers and their contributions, institutional history, as well as the interactions between physics research and society. Also included in the scope of the series are key historical works that are published or translated for the first time, or republished with annotation and analysis.

As a whole, the series helps to demonstrate the key role of physics in shaping the modern world, as well as revealing the often meandering path that led to our current understanding of physics and the cosmos. It upholds the notion expressed by Gerald Holton that "science should treasure its history, that historical scholarship should treasure science, and that the full understanding of each is deficient without the other." The series welcomes equally works by historians of science and contributions from practicing physicists.

These books are aimed primarily at researchers and students in the sciences, history of science, and science studies; but they also provide stimulating reading for philosophers, sociologists and a broader public eager to discover how physics research – and the laws of physics themselves – came to be what they are today.

All publications in the series are peer reviewed. Titles are published as both print- and eBooks. Proposals for publication should be submitted to Dr. Angela Lahee (angela.lahee@springer.com) or one of the series editors.

More information about this series at http://www.springer.com/series/16664

Carl S. Helrich

The Quantum Theory—Origins and Ideas

A Historical Primer for Physics Students

Springer

Carl S. Helrich
Goshen College
Goshen, IN, USA

ISSN 2730-7549 ISSN 2730-7557 (electronic)
History of Physics
ISBN 978-3-030-79267-1 ISBN 978-3-030-79268-8 (eBook)
https://doi.org/10.1007/978-3-030-79268-8

This Springer imprint is published by the registered company Springer Nature Switzerland AG
The registered company address is: Gewerbestrasse 11, 6330 Cham, Switzerland

Preface

> If quantum mechanics hasn't profoundly shocked you, you haven't understood it yet.
>
> Niels Bohr

Any of us who has taught quantum mechanics has tried to convey to our students something of the shock the quantum theory brought to the physics community. This we do in a standard fashion by outlining the experiments that pointed to the failures of classical thinking. This presentation follows the pattern our professors used before they got down to the real objective of the course, which was to present the theory itself.

Of course, our students must also learn the mathematical basis of the quantum theory. Our objective is to show our students how to represent results in terms of sets of eigenfunctions, rather than position and momentum. In my experience, however, students rarely give up the classical sense that position and momentum are properties of the electron. They are normally very interested in the quantum theory and even want to be shocked. But we cannot help them understand how disturbing the twentieth century was if we do not ourselves fully understand what happened.

This book is what I wish had been on my bookshelf as I began each course I taught on quantum mechanics. I was an undergraduate student of engineering at Case when Martin J. Klein was there. So I was familiar enough with his work to realize that there was a lot I did not know. But I also realized that acquiring a fair understanding of the transition in our thinking from classical physics to the quantum theory was going to require an effort, something I kept putting off.

Eventually, I decided to write a text on quantum mechanics with a serious introductory chapter. When that introductory chapter began to get rather long I asked an editor at Springer Nature if this might be acceptable as a book. And as you can see, her answer was affirmative. This book is the result.

My original intention was to provide students with a broader picture of what happened so that they would understand that those critical experiments and ideas did not simply appear from nowhere. Why did the world's leading metrological laboratory focus so much effort on blackbody radiation? And why were electrical discharges in rarefied gases once central in physics? Human personalities also became important as physics developed. Our students should understand this because they are also actors in a continuing drama as we try to understand how things came about.

Because I am writing for students preparing to enter the first course in quantum mechanics, I have assumed the usual level of mathematics. The reader should also be familiar with the classical mechanics and the Maxwell electrodynamics.

Of course, there are also experiments, which are central to the story. I hope to have conveyed something of the experimental genius, dogged hard work, elation, and occasionally even despair that has been so important in the story.

I could see no honest way around beginning with the search for a realistic atomic theory, which had its origins in the thought of the Presocratic Greek philosophers. The atomic structure of matter gradually became a central issue in the eighteenth and nineteenth centuries, and it was behind J. J. Thomson's experiment to identify cathode rays. His primary interest was actually in a rather complex atomic model.

At the other end of the story, the stopping point was relatively easy to identify. A course in quantum mechanics may begin once we have a quantum theory in place. And for that, we normally look to a villa in Arosa in the Swiss Alps where Schrödinger formulated the wave equation in 1926.

Historically, however, there were actually two key discoveries that could serve as stopping points. The first was the matrix formulation of quantum mechanics in Göttingen, in the summer before Schrödinger's winter vacation in the Swiss Alps. It resulted from the work of Born, Heisenberg, and Jordan in Göttingen, whereafter Dirac showed that the difference between the two formulations lies in whether the operators or the basis vectors carry the time dependence.

The matrix mechanics is intimately tied to classical mechanics, with origins in Heisenberg's idea that only observable quantities should be included in the formulation. Schrödinger's first paper, in which the wave equation was developed and solved for the hydrogen atom, relied on the ideas in de Broglie's thesis. The wave and matrix formulations are mathematically equivalent.

Because of the questions that surround Schrödinger's understanding of the wave theory, I have spent some time on his ideas, allowing him to speak through an outline of his Physical Review paper from the end of 1926. However, I give the final word to Born, whose understanding differs fundamentally from Schrödinger's.

I must particularly thank the historian Arianna Borelli and physicist Bretislav Friedrich for reading through this manuscript and suggesting corrections and additions. Their efforts have been invaluable. I would also like to express my gratitude to Angela Lahee the editor at Springer Nature who has worked with me throughout.

Goshen, IN, USA
March 2021

Carl S. Helrich

Contents

1 The Atomic Theory 1
1.1 Introduction 1
1.2 Atoms in Ancient and Early Modern Theories 2
1.2.1 Ancient Theories 2
1.2.2 Early European Thoughts 2
1.2.3 Atomic Motion and Pressure 3
1.2.4 Boscovich's Atomic Model 5
1.3 Interactions 5
1.3.1 Atoms and the Voltaic Pile 5
1.3.2 Phlogiston and Caloric 9
1.3.3 Heat and Thermodynamics 11
1.3.4 Kinetic Theory 17
1.4 Gas Discharges and Cathode Rays 21
1.4.1 Rarefied Gases 22
1.4.2 Cathode Rays 23
1.4.3 Thomson's Experiment 25
1.5 Statistical Mechanics 27
1.5.1 Gibbs' Perspective 27
1.5.2 Ensembles of Systems 27
1.5.3 Statistics of the H-Theorem 29
1.6 Thomson's Atomic Model 29
1.7 Nagaoka's Atomic Model 32
1.8 X-Rays and Electron Number 33
1.8.1 Barkla 33
1.8.2 Diffraction 34
1.8.3 The Braggs 34
1.9 Rutherford and the Nucleus 35
1.10 Summary 39

2 Discovery of the Quantum 41
2.1 Introduction 41
2.2 Planck and the Second Law 42
2.3 Blackbody Radiation 42
2.3.1 Formulae for Radiation Intensity 44
2.3.2 Sources of Radiation 46
2.3.3 Thermodynamics and Radiation 48
2.3.4 Measurements in the Infrared 53
2.4 Probability and the Quantum 58
2.5 Planck's Nobel Lecture 63
2.6 Summary 64

3 Electrodynamics and Matter 67
3.1 Introduction 67
3.2 Planck's Second Theory 68
3.3 Einstein and Photons 69
3.3.1 Regarding One of the Difficulties Encountered by the Theory of "Black Radiation" 70
3.3.2 Regarding the Planck Result for Elementary Quanta 72
3.3.3 Regarding the Entropy of the Radiation 73
3.3.4 Limiting Form of the Entropy for Low Density Monochromatic Radiation 74
3.3.5 Molecular Theoretical Considerations Regarding the Dependence of the Entropy of Gases and Weak Solutions on Volume 75
3.3.6 Interpretation of the Expression for the Dependence of the Entropy of Monochromatic Radiation on Volume According to the Boltzmann Principle 77
3.3.7 Regarding Stokes Law 78
3.3.8 Regarding the Production of Cathode Rays by Illumination of Solid Surfaces 79
3.3.9 Regarding the Ionization of a Gas by Ultraviolet Light 80
3.4 Summary 80

4 Quantum Atoms 81
4.1 Introduction 81
4.2 Niels Bohr 82
4.2.1 Doctorate and Cambridge 82
4.2.2 Bohr's Model 83
4.3 Arnold Sommerfeld 87
4.3.1 Thoughts and Ideas 88

4.3.2 Zeeman Effect 90
4.3.3 Beyond Models 94
4.4 Summary 94

5 Experimental Evidence 95
5.1 Introduction 95
5.2 Bohr Model 95
5.2.1 Ionized Helium 95
5.2.2 Moseley 96
5.2.3 Franck and Hertz 103
5.3 The Einstein Photon 105
5.3.1 Experiments by Millikan 105
5.3.2 Compton's Experiments 109
5.4 Bohr and the Photon 112
5.4.1 Ideas of Bohr, Kramers, and Slater 112
5.4.2 Bothe and Geiger's Experiment 113
5.4.3 Experiment of Compton and Simon 114
5.5 Spatial Quantization: Stern and Gerlach 116
5.6 Summary 121

6 De Broglie's Particle Wave 123
6.1 Introduction 123
6.2 De Broglie's Thesis 125
6.3 The Phase Wave 125
6.3.1 The Relation Between the Quantum and Relativity Theories 125
6.3.2 Phase and Group Velocities 128
6.3.3 Phase Waves in Space-Time 129
6.4 Connections 130
6.4.1 Defense of the Thesis 131
6.4.2 Influences 131
6.5 Davisson–Germer Experiment 131
6.6 Summary 137

7 Göttingen Quantum Theory 139
7.1 Introduction 139
7.2 Background 140
7.3 Matrix Mechanics 145
7.3.1 Kramers and Heisenberg 145
7.3.2 Heisenberg 147
7.3.3 Born and Jordan 151
7.3.4 Born, Heisenberg, and Jordan 159
7.3.5 Enter Dirac 163
7.4 Summary 164

8 Schrödinger's Wave Theory ... 167
8.1 Introduction ... 167
8.2 Schrödinger's First Correspondence ... 168
8.3 Physical Review December 1926 ... 171
8.3.1 An Analogy: Mechanics and Optics ... 172
8.3.2 Analogy and "Undulatory" Mechanics ... 175
8.3.3 Significance of Wavelength ... 175
8.3.4 Application to the Hydrogen Atom ... 177
8.3.5 Discrete Characteristic Frequencies ... 178
8.3.6 Intensity of Emitted Light ... 179
8.3.7 Wave Equation from a Variational Principle ... 180
8.3.8 Physical Meaning of the Wave Equation ... 181
8.3.9 Non-conservative Systems and Dispersion ... 183
8.3.10 Relativity and the Magnetic Field ... 184
8.4 Summary ... 184

9 Spin and Its Interpretation ... 185
9.1 Introduction ... 185
9.2 Spin of the Electron ... 186
9.3 Interpretation of the Wave Function ... 186
9.4 Copenhagen Interpretation ... 190
9.5 Summary ... 194

10 Connecting the Matrix and Wave Theories ... 195
10.1 Introduction ... 195
10.2 Symbolic Method and Representation ... 196
10.2.1 Schrödinger Picture ... 199
10.2.2 Heisenberg Picture ... 201
10.3 Summary ... 203

11 Epilogue ... 205

Appendix A: Hamilton and Fermat ... 207

Appendix B: Schrödinger's Variation ... 213

Appendix C: Schwarz Inequality ... 219

Appendix D: Heisenberg's Principle ... 221

Appendix E: Displacement Operator ... 225

References ... 229

Index ... 237

Chapter 1
The Atomic Theory

A Briton wants emotion in his science, something to raise enthusiasm, something with human interest.

George F. FitzGerald

There can be no descriptive account of the structure of the atom; all such accounts must necessarily be based on classical concepts which no longer apply.

Niels Bohr

1.1 Introduction

The roots of the quantum theory lie in the atomic picture of matter and our attempts to understand the dynamics of atoms, which in turn had their origin in the philosophical problem of existence. The proposal that atoms form the basis of matter did not, however, resolve the fundamental problems facing Presocratic or later philosophers. The laws forming the original basis of physics were discovered without reference to an atomic structure of matter.

Here we highlight some of the steps that were particularly important for developing an atomic theory into a real science consistent with the basic concepts of classical nineteenth century physics. This will take us up to the crucial experiment by Joseph John (J.J.) Thomson (1856–1940), in which he identified the electron as the particle he required for his atomic model. We will then see how Thomson's atomic model failed to explain the X-ray measurements by Max von Laue (1879–1960), and how the experiments by Hans Geiger (1882–1945) and Ernest Marsden (1889–1970) subsequently identified the nucleus.

These results in turn showed that something very fundamental was still missing in our understanding of the physics of the atom.

C. S. Helrich, *The Quantum Theory—Origins and Ideas*, History of Physics,
https://doi.org/10.1007/978-3-030-79268-8_1

1.2 Atoms in Ancient and Early Modern Theories

1.2.1 Ancient Theories

The philosophical problem mentioned in the introduction, that led to the idea of an atomic theory, was Parmenides' (b. ca. 515 BCE) contention that what *is* must be one and unchanging. The difficulty comes from the fact that there can then be no change within what exists, because any change in what is implies that something which previously was not there has come into being. This requires us to speak of "that which is not". But "that which is not" is an unintelligible concept [216].

According to Aristotle, Leucippus (fifth century BCE) tried to escape from this logical dilemma by formulating a theory in which change is only perceived by the senses, but that no "thing" actually changes. He proposed that everything we perceive could be made up of indivisible atoms, from the Greek adjective *atomos* or *atomon*, meaning *indivisible*. His supposed student Democritus (ca. 460–370 BCE) was a firm advocate of this atomic theory. In Leucippus' view, the atoms could neither change nor dissolve into "what is not". They then satisfied the Parmenidean requirement of being. What humans perceived of as change was not change at all in the philosophical sense. When ice melts nothing atomistic changes. It is just that we first perceive a solid and then perceive a liquid [11].

Democritus carried these ideas farther. There were two fundamental realities: atoms and the void. Atoms move about in an infinite void colliding with one another and forming clusters as they connect via barbs and hooks. Perception depends on the action of atoms, and while the atoms were eternal, the products formed from their clustering were not [12].

1.2.2 Early European Thoughts

The works of Aristotle were brought to Europe by the Arabs after their conquest of Spain, and their effect on European thought cannot be exaggerated.[1] Aristotle, who understood that a medium would resist the motion of anything moving within it, produced a mathematical argument denying the existence of the vacuum. This argument led him to declare in *The Physics, Book IV, Section 8* [6] that the vacuum or void cannot exist.

The well known rephrasing of this as *Nature abhors a vacuum* is not a direct translation. Nevertheless, in his *Dialogue on the Two Chief World Systems* [120], Galileo Galilei (1564–1642) used the claim that nature abhors a vacuum in the argument presented by *Salviati*, the character who argues in favor of the Copernican system, to explain the strength of metals. The argument also postulated an extremely large (infinite) number of infinitely small atoms as making up the metal. Galileo then pushed

[1] This is covered extensively in [140, pp. 11–18]. References are contained in that volume.

Aristotle's argument to its logical limit, arriving at the mathematical conclusion that a line is composed of an infinite number of points. This mathematical atomism was primary in Galileo's difficulty with the Jesuits [4, pp. 88–91].

In his *Dialogue* Galileo also noted that water could not be pumped out of a well if the water surface in the well was more than 18 cubits (in modern terms about 10 m) below the rim of the well. In his explanation, Galileo presented a thought experiment, which Salviati explained in terms of the force necessary to break the vacuum, in the same sense as a rope could be broken by an applied force [281].

Then the mathematician and physicist Evangelista Torricelli (1608–1647), who had worked with Galileo just a few months before Galileo's death, decided that the key issue was the weight of the air. In a letter to his friend Michelangelo Ricci (1619–1682), Torricelli pointed out that we live at the bottom of a huge mass of air, which is known to have weight. According to Aristotle, everything except fire had weight. Torricelli made the first barometer by filling a glass tube 2 cubits long with mercury. Then, with his finger over the open end of the tube, he inverted it and placed the open end in a basin containing mercury. The height of the mercury in the tube, up to the beginning of the vacuum at the closed end, was (about) 76 cm in modern units. The pressure holding the column of mercury up, Torricelli reasoned, came from the weight of the air[2] (see [281] and [40, pp. 1,2]).

Blaise Pascal (1623–1662) pointed out that, as fluid pressure increases with depth, the barometric pressure should decrease with elevation. He presumably convinced his brother-in-law Florin Périer (1605–1672) to climb more than 1000 m to the top of the Puy de Dôme, the highest mountain in the vicinity of Clermont-Ferrand where Pascal had been born and raised and where his sister and her husband lived. In September of 1648, Périer carried the necessary equipment to the top of the Puy de Dôme and conducted the Torricelli experiment. He found that the height of the mercury column was 85 mm less than at the base of the mountain.

1.2.3 Atomic Motion and Pressure

Robert Boyle's (1627–1691) celebrated law states that the pressure in a gas is inversely proportional to the volume of the gas at constant temperature, i.e.,

$$P = \frac{F(T)}{V}, \tag{1.1}$$

where $F(T)$ is a function only of the temperature and the amount of gas present. This supposedly led Boyle to a model of the atoms in the gas. However, Gerald Holton (b. 1922) and Stephen Brush (b. 1935) point out that the history of this law is complicated and has only recently been untangled [143, footnote p. 270].

[2] The experiment was actually performed by Vincenzo Viviani (1622–1703), who had also been an assistant to Galileo [281].

Although there is no indication that Robert Boyle (1627–1691) ever claimed that he alone was responsible for this law, his mental picture of the gas was his own. He explained his ideas in detail in a letter dated to 1660. In one of the models, he suggested that the atoms/molecules of the gas could be considered to behave like tiny springs with a tendency to expand. In Boyle's words, the gas was "like a fleece of wool." This was his "static model," which was widely accepted during his lifetime. The second model was his "dynamic model," which no longer treated the identical corpuscles of the gas as stationary, but in a state of agitation, being whirled around by a subtle turbulent fluid.[3] In the static model, a decrease in the volume of the gas compressed the springs, thereby increasing the gas pressure. This was the statement of Boyle's gas law (1.1) in which the pressure of the gas was inversely proportional to the volume at constant temperature T (see [143, pp. 272–273] and [40, pp. 43–51]).

Isaac Newton (1643–1727) discussed forces between gas particles (atoms) in his *Philosophiae Naturalis Principia Mathematica* (*Principia*) (1687), apparently with a view to placing Boyle's law on a firm mathematical footing. In the *Principia*, Newton argued that the pressure would be inversely proportional to the volume, starting from a general force law and a model corresponding to Boyle's. Finally, however, Newton pointed out that the identity of the gas was a physical question that must be settled by experiment (see [40, pp. 52–56] and [143, p. 272]).

In 1738 Daniel Bernoulli (1700–1782) put forward the idea that a gas consists of an enormous number of very small particles moving rapidly and separated by great distances compared to the size of the particles. Based on the fact that gases are compressible, he concluded that most of the gas was empty space. The pressure was the result of collisions of the gas particles with the vessel walls.

In the eighteenth century, many scientists began to speak of heat in terms of motion, but not as free motion of the particle constituents of matter. Bernoulli's model in which the gas particles were in rapid translational motion, asserting that the pressure on the walls of the vessel was not a constant force but the result of a large number of impacts of extremely small particles, was completely new. With this model of a gas, Bernoulli was able to produce the inverse volume $1/V$ factor in Boyle's Law (1.1) (see [40, pp. 57–65], [217, p. 253], and [143, p. 273]). Bernoulli's ideas still constitute our basic picture of the ideal gas, which emerged with James Clerk Maxwell (1831–1879) and Ludwig Boltzmann (1844–1906) in the nineteenth century.

As with many theories that are far ahead of contemporary thought, Bernoulli's gas theory actually had little effect on the development of atomic theory in the eighteenth century. It simply did not come to grips with the problems that were considered primary. Among these were the complexities of the motion of atoms through the aether and the interactions of atoms with one another. Bernoulli had also dispensed with discussions of heat at a time when the caloric theory of heat was accepted. He identified temperature with the kinetic energy of the atoms, but this did not come to grips with the critical question of heat as it was then posed [40, pp. 7,8].

---

[3] Holton and Brush remark that this was the beginning of a futile 250 year search for an ethereal fluid [143, p. 272].

Brush astutely points out that the person who persuades the world to adopt a new idea has accomplished as much as the person who conceived the idea [40, p. 9].

1.2.4 Boscovich's Atomic Model

In his work *Theoria Philosophiae Naturalis* (1758), the Jesuit Ruggero Giuseppe Boscovich (1711–1787) presented his ideas on what we may consider to be the first coherent description of an atomic theory ever published.

Through his atomic theory, Boscovich attempted to comprehend the structure of matter in a single idea. His concept of the physical atom went beyond the previous ideas of extended hard and elastic bodies. He postulated atoms to be point particles surrounded by fields of attractive and repulsive forces. The forces were repulsive at very small distances, becoming attractive or repulsive alternately with increasing distance, and finally becoming attractive, following Newton's law of gravitation, at macroscopic distances.

He considered a chemical element to be made up of a number of these points with rather complicated fields surrounding them. Molecules were then formed from elements, as was already known in chemistry. Boscovich proposed that the resulting structure of these molecules was such that they could be subjected to considerable force without becoming dissociated.

Boscovich's atomic theory raised questions concerning the reality of continuous lines and geometrical space. We already noted that these mathematical questions were at the heart of Galileo's conflict with the Jesuits. His ideas on continuity can be found in a treatise held in the Catholic University at Brescia. Apparently Boscovich, a Jesuit priest, was able to consistently argue for the continuity of the line and also of solid matter by exploiting the point character of his atoms with their fields of force. In the eighteenth and nineteenth centuries, his atomic theory had considerable influence on scientific thought, particularly in Britain [38, 145, 176, 276].

1.3 Interactions

1.3.1 Atoms and the Voltaic Pile

Apart from the explanation of Boyle's law based on Bernoulli's model of a gas, there were no laboratory experiments revealing the atomic structure of matter in the mid-eighteenth century. This changed with the growing interest in electric and magnetic phenomena.

Beginning in 1780, Luigi Galvani (1737–1798), professor of anatomy at Bologna, experimented with the response of frogs' legs to electrical excitation. He attributed

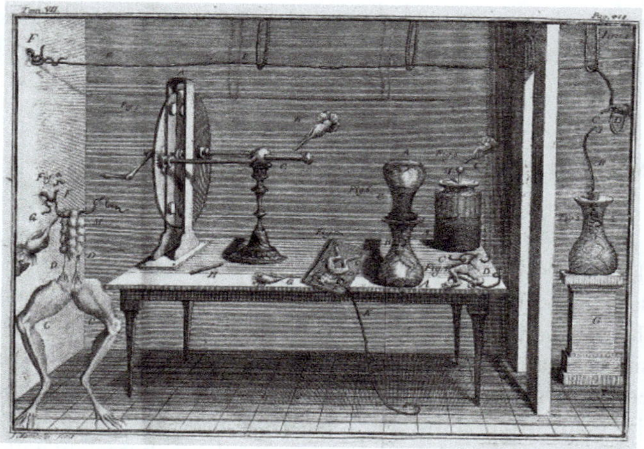

Fig. 1.1 Luigi Galvani's experiment with frog legs (1786). [The author died in 1798, so this work is in the public domain in its country of origin and other countries and areas where the copyright term is the author's life plus 100 years or fewer.]

Fig. 1.2 Library of the University of Bologna circa 1918. [Author: Unknown. Source: Hector Buissneg. Public Domain]

the response to the transfer of a fluid between the nerves and the muscles. Figure 1.1 is a print from 1791 showing Galvani's experiment.

Although Galvani thought that what was happening in a frog's leg was no different from the discharge of a Leyden jar, he had captured human imagination with his idea of animal electricity [289, pp. 69–70] (Fig. 1.2).

Alessandro Volta (1745–1827), professor of natural philosophy at the University of Pavia, thought that the leg of the dead frog was of secondary importance and that the moist tissues of the frog's leg were simply incidental. In 1793 he was prepared to completely abandon any reference to animal electricity and by 1799 he was developing the idea of connecting together sandwich-like combinations of dissimilar metals and moist materials to produce electricity. This led to the voltaic cell or pile (see, e.g., [139, p. 8] and [289, p. 70]).

Then a particularly important step was taken by Giovanni Fabroni of Florence (1752–1822). Fabroni dispensed with the medium supporting the moisture and simply used water. He observed that one of the plates submerged in the water became slightly oxidized when the plates were later brought into contact and concluded that some kind of chemical reaction involving the plates and the water must be important [289, p. 71].

Volta announced his discovery of the pile in a letter to the President of the Royal Society of London, Sir Joseph Banks (1743–1820), in March of 1800. Banks passed this information on to William Nicholson (1753–1815) and Anthony Carlisle (1768–1840). Nicholson and Carlisle had a voltaic pile on their laboratory bench before the end of April.

To secure one of the contacts, Nicholson and Carlisle added a drop of water to it and noticed a gas bubbling off. They then stuck both terminals of the pile into the ends of a tube filled with water. At one end a flammable gas was produced, while the other end became oxidized. When they used platinum wires, they produced free oxygen at one end and free hydrogen at the other.

The precise identity or definition of electricity was unclear. However, the connection between electricity and chemical reactions was becoming clear, and by the end of the eighteenth century the formulation of chemistry in terms of atoms was emerging. In 1758 Boscovich had already proposed a complex and stable structure for molecules that were built up from atoms (see Sect. 1.2.4). Antoine Lavoisier (1743–1794) subsequently presented a law of conservation of mass for chemical reactions (see Sect. 1.3.2.1) [147, p. 185].

The work of Nicholson and Carlisle on the voltaic pile attracted the interest of Humphry Davy (1778–1829), who had just been appointed professor of chemistry at the Royal Institution in London. Davy began experimenting with voltaic piles in November of 1800 and noticed that there was no voltage between the poles of the pile if the liquid separating them was distilled water. The liquid had to be a conductor. And he observed that the liquid had also to be capable of undergoing chemical reactions with the poles. He then concluded that the chemical reaction (at the poles) was the source of the electrical effects. Nicholson suggested that these electrical effects required the use of different metals for the poles [289, pp. 75–77].

Figure 1.3 is a painting by Thomas Hosmer Shepherd (1793–1864) of the Royal Institution as it appeared in around 1838.

Looking at the flow of current through a liquid, most investigators considered the poles to be of primary importance. In 1811, however, Joseph Louis Gay-Lussac (1778–1850) and Louis Jacques Thénard (1777–1857) suggested that the rate of decomposition of the electrolyte might depend only on the total current flowing through the electrolyte and that the size of the electrodes and the strength of the electrolyte were of no consequence.

Fig. 1.3 Painting of the Royal Institution of Great Britain in London, circa 1838. Painted by Thomas Hosmer Shepherd (1793–1864) [Public Domain]

Davy also suggested that a chain of decompositions and recombinations might occur in the liquid and that the decomposition resulting in the gases appearing at the terminals might be due solely to the forces at the terminals. Those forces depended on the identity of the metal in the terminal.

In 1833 Michael Faraday (1791–1867), who had just been appointed first Fullerian Professor of Chemistry at the Royal Institution in London, designed an experiment to test this proposal. He soaked paper in salt solution and mounted it with wax between the separated terminals of an electrostatic machine. The salt in the paper was decomposed, although there was no contact with the terminals. The terminal connection was thus irrelevant. The salt in the paper was decomposed by separation of its positive and negative ions and the motion of these ions constituted the current. The metal terminals served only to stop this flow [289, pp. 197–198].

Faraday called the terminals electrodes, identifying the positive electrode as the anode and the negative electrode as the cathode. Then anions (such as Cl^-) were produced at the anode and cations (such as Na^+) were produced at the cathode. The material that was decomposed was called the electrolyte. The electrolyte decomposed into anions and cations according to the electrode at which each was produced [289, p. 199].

Davy and Faraday both thought in atomic terms. They had been influenced by the ideas of Boscovich, which we discussed in Sect. 1.2.4. Faraday's concept of the field may have been rooted in the ideas of Boscovich (see [38] and [289, p. 199]).

Faraday spoke in terms of the chemical equivalent, which was the amount of a substance required to combine with some standard amount of a certain element. If the element is hydrogen the chemical equivalent is related to valence and the atomic mass.

His experiments with electrolytes allowed Faraday to conclude that the atoms of matter are endowed with electrical properties. This electrical property determined the binding force between the anion and cation of the electrolyte. The voltage at which the electric current was produced depended on the energy of the chemical reaction. This could be measured directly.

Faraday admitted that it was very easy to speak of atoms but more difficult to form an idea of their nature [289, p. 201].

1.3.2 Phlogiston and Caloric

The experiments by Davy and Faraday provided insights into the electrical and chemical aspects of voltaic piles. The basic formulation of the electrical force between charges, based on experiment, had been published by Charles Augustin Coulomb (1736–1806) in 1785 [289, p. 58]. A more complete understanding of the voltaic pile required a unified formulation of the chemistry and physics at an atomic level. For this to be possible, observations of the behavior of matter had to be cast in terms of the dynamic behavior of atoms. But two concepts stood in the way: phlogiston and caloric.

1.3.2.1 Phlogiston

The phlogiston theory was proposed in 1667 by Johann Joachim Becher (1635–1682) and developed by Georg Ernst Stahl (1660–1734). An element, which was called phlogiston, was carried off in the flames, or light, during combustion. Although incorrect, phlogiston theory provided the first great generalization in chemistry. Phlogiston could be used to explain chemical changes involving elements.[4] Like electricity, however, phlogiston could not be put in a flask and labeled.

In experiments performed over the relatively short time span of two years (1772–1774), Antoine Lavoisier (1743–1794) found that sulphur and phosphorus gained rather than lost weight on burning and that the total weight of air plus the material being burned in a closed vessel did not change. Lavoisier had established the law of mass conservation in a reaction. If two interacting molecules with masses m_1 and m_2 form two molecules with masses m_3 and m_4 in a chemical reaction then [147, p. 185]

$$m_1 + m_2 = m_3 + m_4.$$

Although these experiments were convincing, Lavoisier did not immediately declare the end of the phlogiston idea. He waited until after his experiments of 1782–

[4] Boyle had defined an element as "a substance that cannot be decomposed into any simpler substance."

1783 with Pierre Simon, Marquis de Laplace (1749–1827), which later formed the basis of thermochemistry. Then he announced that the phlogiston picture was unnecessary and replaced it by a simpler explanation based on experimental evidence. [D. McKie: Introduction to [169], pp. ix–xxiv].

With the elimination of phlogiston as some kind of subtle fluid, and the pronouncement of mass conservation as a law that must hold in all chemical reactions, science had come closer to an atomic theory of matter. But there were still questions regarding the behavior of matter when heated or cooled that had to be answered.

1.3.2.2 Caloric

In the nineteenth century, caloric theory was accepted as an explanation for heat and its effect on matter. According to Lavoisier it was difficult to understand any of the effects of heating and cooling of substances without admitting that they were the result of the action of a real and material substance, or very subtle fluid, separating the particles of bodies from one another. The concept of caloric first appeared on the list of elements in the Chemical Nomenclature worked out in 1787 by Guyton de Morveau (1737–1816), Lavoisier, Claude Louis Berthollet (1748–1822), and Antoine François de Fourcroy (1755–1809). The concept of caloric was then adopted by most English chemists (see [169, pp. 4, 5] and [40, pp. 11, 12]).

Caloric was the repulsive cause for the separation of particles in matter under heating. The result was an expansion of the matter and formation of first a liquid and then a gas when it became hot enough. The pressure of the atmosphere kept liquids from all becoming gases as a result of the expansive force of caloric. Because caloric penetrated all matter, no vessel could contain it. Free caloric could exist, but because of its strong adhesive properties, it could never be obtained in the free form.

The differences in heat capacities of materials could be easily understood in terms of the attractive forces between the particles of matter. This force caused equal weights of substances to differ in the amount of caloric required for them to reach the same temperature.

Sensible heat was the effect on our sense organs made by the passage of caloric. Lavoisier put forward the axiom that without motion there would be no sensation.

As Brush points out, there was little reason for the scientists of the nineteenth century to question the validity of the caloric theory [40, p. 9].

Laplace proposed a mechanical theory of heat in 1783, despite Lavoisier's objections. But then Laplace changed his mind and became an advocate of caloric theory [256, p. 216]. Newton had made no attempt to explain the repulsive forces between the particles of a gas. Laplace, however, in his *Traité de Mécanique Céleste*, Book XII (Paris 1825) presented a caloric-based theory for the properties of gases. He postulated a repulsive force between any pair of similar gas particles (molecules), each containing an amount of caloric c. This repulsive force between like particles separated by a distance r was of the form

$$F = Hc^2\phi(r), \tag{1.2}$$

where H is a constant and $\phi(r)$ is a rapidly decreasing function of r. The value of the constant H could depend on the gas. Laplace argued that the amount of caloric c surrounding the atom must be a function of the gas density. He assumed that the attractive force operable over greater distances was gravitational in origin.

Each atom sent out and received rays of caloric at a rate dependent on the temperature,[5] which Laplace denoted by u. His final analysis yielded the function

$$P = i\rho\Pi(u), \tag{1.3}$$

for the gas pressure P. Here ρ is the gas density, i is a constant (not the imaginary unit), and $\Pi(u)$ is a function of the temperature u alone.

Laplace defined $\Pi(u) = \rho c^2/q'$, where the value of the constant q' depends on the gas. Equation (1.3) is the ideal gas law, provided that $\Pi(u) = u$. Laplace contended that this was the case because $\Pi(u)$ is a measure of the density of radiant caloric in space. These ideas appeared in a paper read to the Royal Academy of Sciences in Paris in September of 1821 [40, pp. 11–13].

Laplace's use of the caloric theory was not without success. He obtained the representation of adiabatic changes of the state of gases as $PV^{\gamma} = \text{constant}$, where γ is the ratio of the specific heat at constant pressure to that at constant volume. Assuming that sound waves involved an adiabatic compression, he was also able to correct Newton's result for the speed of sound by adding a factor $\sqrt{\gamma}$, which resulted in good agreement with experiment. Laplace's model predicted, however, that the specific heat of a gas should increase with decreasing pressure or density. Brush points out that, if experimental techniques had been better in the early nineteenth century, the caloric theory might have been abandoned much sooner [40, p. 13].

Many scientists already accepted that there was a connection between particle motion and heat. The transition from solid to liquid to gas implies a greater state of motion under heating. So the concept of heat was not easily related only to the presence of caloric. But scientists were not prepared to accept that heat was only particle motion [40, p. 14].

1.3.3 Heat and Thermodynamics

The complexities in the universality of caloric made the theory difficult to unseat by direct theoretical arguments. The path to removing the theory began with attempts to produce useful work from heat and the observation that doing work on a body increases its temperature.[6]

---

[5] Temperature was not yet understood in thermodynamic terms. We therefore use Laplace's designation u.

[6] Benjamin Thompson (Reichsgraf von Rumford, or in English, Count Rumford) (1753–1814) conducted experiments on cannon barrels, while in the service of the Duke of Bavaria. At the arsenal in Munich in 1798, Thompson mounted a cannon barrel vertically in water and and arranged for

1.3.3.1 Carnot's Ideas

The use of heat to produce motive power was initially the result of the ingenuity of inventors. There was no science involved until the young French military engineer Sadi Carnot (1796–1832) published a remarkable memoir *Réflexions sur la puissance motrice du feu et sur les machines propres à développer cette puissance* (Reflections on the Motive Power of Fire) in 1824. Carnot used caloric in this memoir, although his notes show that he later rejected the concept of caloric. In the memoir Carnot made three main points:

1. Heat engines work because caloric flows through them much as water flows over a water wheel.
2. An efficient heat engine should operate on a cycle, in which the working substance[7] remains enclosed in a cylinder fitted with a piston.
3. The most efficient engine should be reversible in the sense that in each step there is a very small difference between the forces and temperatures within the cylinder of the engine and those outside. Very small changes in external conditions could then reverse the direction of the process and the heat engine could operate as a heat pump, pumping the caloric upwards across the temperature difference [45].

Figure 1.4 shows a symbolic drawing of Carnot's idea for the most efficient engine cycle. We have used heat rather than caloric in Fig. 1.4. Carnot's cycle converts heat into work. In his paper, written in terms of caloric, which was conserved, work could not appear as a conversion of caloric. Figure 1.4 would then have had separate arrows for the work produced and the caloric flow.

Because temperature differences meant loss of efficiency, heat was transferred only under conditions of constant temperature (T_1 and T_2). Other processes were adiabatic (no heat transfer).

Carnot firmly believed there could be no perpetual motion. This allowed him to prove that the efficiency of his cycle operating between two temperatures would be identical for all working substances. The critical point was that the working substance in the cylinder was irrelevant.

Carnot's theorem is now taken to state that $\sum_i Q_i/T_i = 0$, where Q_i is the heat transferred to or from the cycle in the ith leg of the Carnot cycle at the thermodynamic temperature T_i. However, Carnot never managed to show that the function he called $F(t)$ was $1/t$, where t was the temperature in degrees Celsius. In 1848 William Thomson (Lord Kelvin) (1824–1907) first proposed a thermodynamic temperature T, which could be defined in terms of the efficiency of the Carnot cycle.

Carnot's theorem would lead to the definition of entropy by Rudolf Clausius (1822–1888) (see [35, p. 241], [287, p. 21], [143, pp. 256–257], [40, p. 16], and [271]).

---

the barrel to be bored by rotating a blunted boring tool concentric with the barrel. The water was brought to a boil in two and a half hours.

Thompson's experiments were taken seriously in France, although ignored in England. The measurements did not produce satisfactory (scientific) data. The link between motion and heat, however, could not be easily denied [256, p. 215].

[7] Carnot used air in his example.

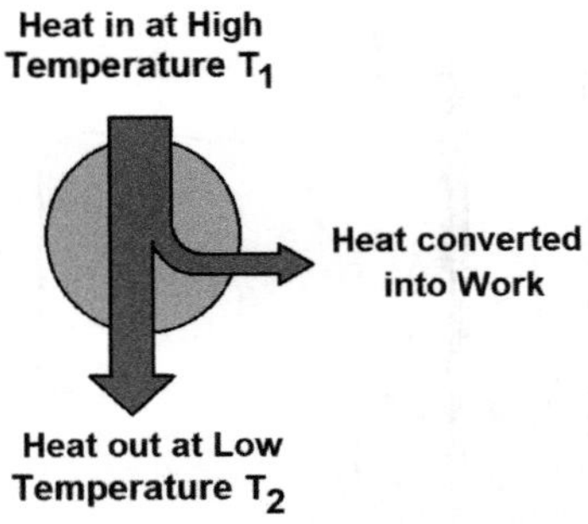

Fig. 1.4 Carnot's most efficient conversion of heat into work. Heat enters from the high temperature reservoir only while the cylinder temperature is equal to the reservoir temperature and exits to the low temperature reservoir only when the temperatures of the cylinder and the low temperature reservoir are equal. The other legs in the cycle involve no heat transfer. [Drawn by CSH.]

Carnot also provided the modern definition of thermodynamic work as the raising or lowering of a weight in the gravitational field. Although Carnot was mathematically competent, he wanted his memoir to be read and understood by engineers, so he avoided mathematics. Unfortunately, the memoir was essentially ignored by both engineers and scientists until its republication by Émile Clapeyron[8] two years after Carnot's death. Carnot died of cholera during the 1832 epidemic in Paris.

1.3.3.2 Joule's Experiments

The experiments that finally removed caloric from serious consideration were those of the amateur scientist James Prescott Joule (1818–1889). Joule was able to show that the temperature of a fluid in an insulated vessel could be increased by mechanical work done on the fluid. He presented his results at the meeting of the British Association for the Advancement of Science in Oxford in 1847. That Joule's paper was noticed at all was a result of the fact that Thomson had pointed out the importance of the experiment (see [143, pp. 243, 256], [63, pp. 64, 65], and [138, p. 5]).

Figure 1.5 shows a drawing of Joule's setup for the experiment that attracted Thomson's attention. The falling weight produced thermodynamic work on the paddle wheel inside the insulated vessel containing water. Only work was then transferred.

The experiment illustrated in Fig. 1.5 should not be considered as Joule's only experiment. He began these experiments in 1840 and many of them were technically very difficult. To measure the temperature rise, he had thermometers capable of measuring a hundredth of a degree Fahrenheit made by John B. Dancer (1812–1887), an instrument maker in Manchester [63, p. 64].

[8] French engineer Benoît Paul Émile Clapeyron (1799–1864).

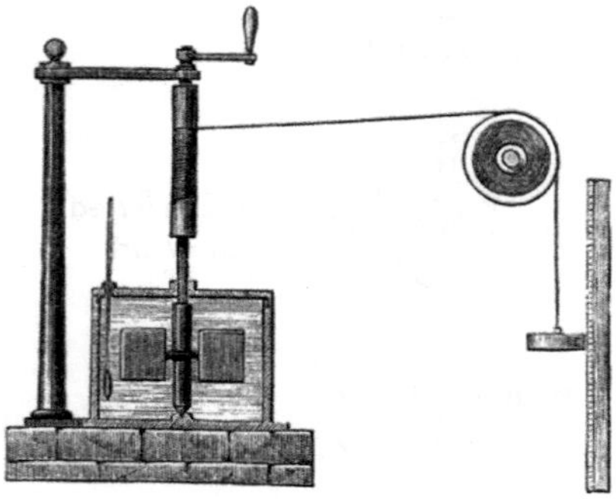

Fig. 1.5 Joule's apparatus for one of his experiments. This is the experiment normally cited and is the one that attracted Thomson's attention. [This drawing first appeared in *Harper's New Monthly Magazine*, No. 231, August, 1869. This work is in the public domain in the United States because it was published (or registered with the U.S. Copyright Office) before January 1, 1925.]

The importance of these experiments soon began to get noticed. The French journal *Comptes Rendus* published a short account in 1847 and Joule was elected a corresponding member of the Royal Academy of Sciences in Turin in 1848. In 1850 he was elected a Fellow of the Royal Society of London [63, p. 65].

1.3.3.3 Helmholtz and Energy Conservation

Although the subject matter treated did not yet include molecules, the science of mechanics had reached a high point by 1837. In the hands of William Rowan Hamilton (1805–1865) and Carl Gustav Jacobi (1804–1851), the analytical mechanics of Pierre Louis Maupertuis (1698–1759), Leonhard Euler (1707–1783), and Joseph-Louis Lagrange (1736–1813) had been cast in an elegant closed form based on a single function. This function S was known as Hamilton's *principal function*.

The principal function was expressed in terms of the functions T and U, where $2T$ was the *vis viva* or *living force* and U was the *force function*. The functions T and V $(= -U)$ are now known as the kinetic and potential energies [140, pp. 31–46]. The status of theoretical mechanics will be important in the following discussion.

In 1847 Hermann von Helmholtz (1821–1894) published *Über die Erhaltung der Kraft* (in English, *On the conservation of energy*). In the Introduction Helmholtz wrote: "The final goal of theoretical sciences is to find the final immutable causes of what occurs in nature." He devoted the first sections to analytical mechanics with forces (modern terminology) dependent on distance. In section IV he treated the equality of energy and heat. And in sections V and VI he considered electrical and the electromagnetic forces. He concluded that the conservation of force (energy) applies in all cases (see [137, p. 110] and [40, p. 110]).

1.3.3.4 Clausius and the First Law

The classical statement of the first law of thermodynamics followed from Joule's experiment with Carnot's definition of thermodynamic work as the raising or lowering of a weight in the gravitational field. By convention, positive work is work done by the system. Heat is then defined as the difference between the work done by an adiabatically enclosed system and by a diathermally enclosed system between the same states. Heat entering the system is taken is positive.

In his paper of 1850 Rudolf Clausius (1822–1888) clearly stated that the assumptions of the caloric theory were false and presented the first law of thermodynamics, based on Joule's experiments, as

$$\delta Q = \mathrm{d}U + \delta W, \tag{1.4}$$

where U is the internal energy of the system, δQ is the heat transferred, and δW is the thermodynamic work done. The symbol δ indicates that the quantity is not an exact differential. Clausius and Thomson did not agree over Clausius' understanding of energy pathways within the system. However, in 1851 Thomson wrote that it was indeed Clausius who had established Carnot's theorem on the basis of correct principles [108]. All differences were resolved in 1865 [63, pp. 96–97].

1.3.3.5 Clausius and the Second Law

In 1854 Clausius went beyond the first law. He accepted that, as heat passed through a cycle operating between two heat reservoirs at different temperatures, a portion of the heat simply passed from the higher to the lower temperature, while some of it was used to produce work (see Fig. 1.4). No cyclic process, however, could transmit heat from a low temperature to a higher temperature without having work done on it. This was Clausius' statement of the second law of thermodynamics.

Clausius' development of the mathematical statement of the second law in 1854 had roots in Carnot's work, and his thoughts continued to develop over the following 15 years. Mathematically, Carnot's theorem states that $\sum_i Q_i/T_i = 0$, where the subscript i indicates the process (isothermal or adiabatic) in the Carnot cycle. Since any reversible cycle can be constructed as the sum over an arbitrary number of (reversible) infinitesimal Carnot cycles, Clausius used Carnot's theorem to obtain

$$\oint \frac{\delta Q_{\mathrm{rev}}}{T} = 0, \tag{1.5}$$

for any reversible cycle. We have included the subscript rev on the heat transfer term to indicate that the cycle is reversible. Mathematically, (1.5) defines a property S of the system using the differential $\mathrm{d}S = \delta Q_{\mathrm{rev}}/T$.

If any part of the cycle is irreversible, the equality in (1.5) becomes an inequality

$$\oint \frac{\delta Q}{T} < 0. \tag{1.6}$$

The direction of the inequality sign is a result of the fact that any irreversiblility will result in a loss of heat from the cycle that could otherwise have been used to produce work [287, pp. 22, 23].

To determine the effect of a spontaneous, irreversible process on the property S of the system, we consider that such a process takes place in an isolated system and carries the system from state A to state B. We then return the system to the state A by a reversible process for which we can calculate the change in S. Hence, for the total cycle going spontaneously from A to B and returning reversibly to A, we have

$$\int_{\mathrm{A}}^{\mathrm{B}} \frac{\delta Q}{T} + \int_{\mathrm{B}}^{\mathrm{A}} \frac{\delta Q_{\mathrm{rev}}}{T} < 0. \tag{1.7}$$

Since $\mathrm{d}S = \delta Q_{\mathrm{rev}}/T$, (1.7) becomes

$$\int_{\mathrm{A}}^{\mathrm{B}} \frac{\delta Q}{T} + S_{\mathrm{A}} - S_{\mathrm{B}} < 0. \tag{1.8}$$

Since the system was isolated when the spontaneous process occurred, $\delta Q = 0$ and

$$S_{\mathrm{A}} < S_{\mathrm{B}}. \tag{1.9}$$

A spontaneous irreversible process occurring in an isolated system results, therefore, in an increase in the function S.

We thus have a general formulation of the integral (1.6) as

$$\oint \frac{\delta Q}{T} \leq 0. \tag{1.10}$$

This is Clausius' inequality, which is now the mathematical statement of the second law.

1.3.3.6 Entropy

In 1865 Clausius identified the property S as the *entropy*. The word comes from the Greek $\eta\tau\rho o\pi\eta$, which means transformation [158, p. 80]. Clausius' demonstration that entropy must increase as the result of an irreversible (spontaneous) process in an isolated system was one of the great steps in theoretical physics. The entropy is then a measure of the irreversible nature of a process occurring in a system. As Clausius wrote in 1865, the energy of the world (universe) remains constant. The entropy always goes toward a maximum [54].

1.3.3.7 Verbal Statements

In words, the second law of thermodynamics is a negative statement. There are two equivalent formulations. Lord Kelvin's version states that no cycle can convert all the heat transferred to it from a reservoir at a single temperature into work. As noted above, the Clausius version states that no cycle can transfer heat from a low to a high temperature reservoir without work being done on the cycle (see [63, pp. 98–102], [287, pp. 21–23], and [138, p. 8]).

1.3.4 Kinetic Theory

As James Jeans (1877–1946) pointed out, after Bernoulli's contribution in 1738 there was almost no work on kinetic gas theory for a century [151, p. 3]. With some exceptions this later effort in kinetic theory followed the establishment of the first and second laws of thermodynamics. The notable exceptions are the contributions of John Herapath (1790–1868)[9] and John James Waterston (1811–1883),[10] along with Laplace's work on atomic forces in 1825 (see Sect. 1.3.2.2) [41, pp. 5–9].

The first law of thermodynamics was rather easily understood in terms of atomic motion, along the lines of Bernoulli's proposal. The total energy is conserved for a collection of mechanical atoms obeying Newton's laws. The second law, however, brought to bear an immediate difficulty. A time directionality is imposed by the requirement that entropy must always increase.[11] The laws of classical mechanics were time reversible. Reversing the vector momentum of an atom would result in the atom's retracing its trajectory exactly. If the kinetic theory of atoms could not describe the irreversibility of the second law, there was reason to reject the atomic theory.

1.3.4.1 Clausius

Clausius first treated heat as motion in a detailed manner in 1857. He acknowledged the earlier ideas of August Krönig (1822–1879), which were similar [164]. They agreed on motion of atoms between collisions, but Clausius acknowledged rotation and vibration of molecules, which could be transferred on collision.[12] Moreover,

---

[9] Herapath's paper was initially rejected by the Royal Society for publication in its *Philosophical Transactions*. It was published in the *Annals of Philosophy* in 1821, which merged with the *Philosophical Magazine* in 1826.

[10] Waterston's contributions to the kinetic theory suffered a similar fate to that of Herapath's. He eventually presented his work to the British Association for the Advancement of Science in 1851.

[11] For a system out of equilibrium, the entropy change includes an internal change, which the second law requires always to be greater than zero [138, Sect. 11.2].

[12] Krönig served as editor of *Die Fortschritte der Physik* and had at least encountered the abstract of Waterston's 1851 paper. Daub discusses the influence of Waterston on Krönig [65].

Clausius did not reject the idea that some finer kind of matter might be moving with the atomic masses without separating from them. The treatment Clausius presented was primarily verbal and considered liquid to vapor transitions. His mathematical treatment provided an understanding of the specific heats of a gas as well as their ratio [50, 51].

In 1858 Christophorus Buijs-Ballot (1817–1890) objected to Clausius' calculation of the velocity of atoms in a gas. He pointed out that if the molecules of gas moved in straight lines, volumes of gases in contact would mix very rapidly, but this is not observed. As an example, he pointed out that: "If sulphuretted hydrogen or chlorine be evolved in one corner of a room, entire minutes elapse before they are smelt in another corner" [43].

In response Clausius developed the concept of the mean free path, which estimates the distance an atom will travel between consecutive collisions. The effective volume of a gas atom must also include its range of influence in collisions. This is greater than the volume of the atom itself [52, 53].

1.3.4.2 Maxwell

Maxwell picked up the question of the motion of atoms in a gas in 1860. He discussed collisions in general and noted that in a short amount of time collisions would cause the kinetic energy (vis viva) to be distributed among the atoms according to some regular law.

Although Maxwell denoted the velocity components by x, y, and z, we shall use the modern notation v_x, v_y, and v_z to avoid any confusion. He identified the number of atoms (particles) with velocity in the x-direction between v_x and v_x+dv_x in terms of a distribution function $f(v_x)$ as $Nf(v_x)\mathrm{d}v_x$, where N is the number of atoms (particles) present. With no preferred direction, he was able to show that the distribution function for a particular velocity component is

$$Nf(v_\mathrm{j}) = N\frac{1}{\alpha\sqrt{\pi}}\exp\left(-\frac{v_\mathrm{j}^2}{\alpha^2}\right). \tag{1.11}$$

And for the number of atoms for which the magnitude of the velocity lies between v and v+dv, integrating over all directions yields

$$Nf(v) = N\frac{4}{\alpha^3\sqrt{\pi}}v^2\exp\left(-\frac{v^2}{\alpha^2}\right). \tag{1.12}$$

The importance of Maxwell's distribution function and the approach Maxwell introduced for the study of gas dynamics cannot be overestimated. Bernoulli, Krönig, and Clausius had assumed that the velocities of all the atoms (molecules) would be identical. Maxwell's value for the mean free path had differed slightly from the one

Clausius had used, but Clausius subsequently agreed with Maxwell [40, footnote pp. 160–161].

1.3.4.3 Boltzmann

Ludwig Boltzmann (1844–1906) picked up the methodology laid out by Maxwell. Specifically, Maxwell's memoir of 1866 contains a description of the change in the distribution function $f(\mathbf{v})$ as a result of collisions between atoms (molecules) on which Boltzmann relied in formulating the collision term in his equation, known as the Boltzmann equation [182]. Since he was interested in the kinetic basis of gas dynamics, Boltzmann took the distribution function to be dependent on space and time as well as velocity. The distribution function then became $f(\mathbf{r}, \mathbf{v}, t)$ and the partial differential equation for $f(\mathbf{r}, \mathbf{v}, t)$ became, for a single component gas,

$$\begin{aligned} &\left(\frac{\partial}{\partial t} + v_x\frac{\partial}{\partial x} + v_y\frac{\partial}{\partial y} + v_z\frac{\partial}{\partial z}\right. \\ &\left. +F_x\frac{\partial}{\partial v_x} + F_y\frac{\partial}{\partial v_y} + F_z\frac{\partial}{\partial v_z}\right) f(\mathbf{r}, \mathbf{v}, t) \\ &= \left(\frac{\partial f}{\partial t}\right)_{\text{coll}}, \end{aligned} \tag{1.13}$$

with

$$\begin{aligned} \left(\frac{\partial f}{\partial t}\right)_{\text{coll}} = \int_b\int_\varepsilon\int_{v_1} &\left[f\left(\mathbf{r}, \mathbf{v}', t\right) f\left(\mathbf{r}, \mathbf{v}_1', t\right)\right. \\ &\left. - f(\mathbf{r}, \mathbf{v}, t)\, f(\mathbf{r}, \mathbf{v}_1, t)\right] g b \mathrm{d}b \mathrm{d}\varepsilon \mathrm{d}\mathbf{v}_1 , \end{aligned} \tag{1.14}$$

where m is the atomic (molecular) mass and F_j the jth component of a possible body force on the atom (gravity). The term (1.14) is the *collision term*, which provides the rate of change of $f(\mathbf{r}, \mathbf{v}, t)$ due to binary collisions between molecules of the same kind.[13]

This collision term contains what is known as the assumption of molecular chaos.[14] This is the assumption that the probability of the simultaneous occurrence of two atoms with velocities $\mathbf{v}$ and $\mathbf{v}_1$ in a differential spatial volume $\mathrm{d}\mathbf{r}$ around $\mathbf{r}$ is equal

---

[13] The primes in the collision term distinguish properties after collision from those before, g is the relative velocity of the atoms before the collision, b is the impact parameter, and ε places the plane of collision in space. The subscript 1 indicates the colliding partner. The species of the colliding partners in (1.14) are the same. For multi-component gases there is a collision term for each component [22, p. 123].

[14] This was originally called the *Stosszahlansatz* or collision rate assumption. It was considered self-evident by both Maxwell and Boltzmann. The postulate of molecular chaos replaced the *Stosszahlansatz*. Modern studies in the kinetic theory of gases represent the phase space density as a hierarchy in the so-called BBGKY expansion (see, e.g., [199, pp. 42–49] and [235, p. 286]). The Boltzmann equation is the first order approximation to the BBGKY expansion.

to the product of their probabilities of being in that volume, i.e., $f(\mathbf{r}, \mathbf{v}, t) f(\mathbf{r}, \mathbf{v}_1, t)$. The hypothesis of molecular chaos is basically the assumption that the velocities of the various atoms in the gas are not correlated [48, p. 58].

From Eq. (1.13) Boltzmann obtained, for example, the equations of fluid dynamics and of heat conduction [22, pp. 141–197]. Our present kinetic theory of gases is based on the Boltzmann equation (see Mintzer in [177]). Starting from the collision term in (1.13), Boltzmann was able to show that the function

$$H(t) = \int \mathrm{d}\mathbf{v} \left[f(\mathbf{v}, t) \ln f(\mathbf{v}, t) \right] \tag{1.15}$$

satisfies $\mathrm{d}H(t)/\mathrm{d}t \leq 0$ and that $\mathrm{d}H(t)/\mathrm{d}t = 0$ only if

$$f(t) \propto \exp\left(-mv^2/2kT\right) ,$$

where m is mass of the gas atoms and $k =$ constant. This is Boltzmann's H-theorem and $\exp\left(-mv^2/2kT\right)$ is the Maxwellian distribution (1.11).

This was a remarkable discovery, but it came under immediate criticism, primarily because the Newtonian laws of mechanics on which it was based were reversible in time. It was, therefore, logically impossible to obtain an irreversible statement from those laws [22, pp. 412–446].

Boltzmann found the criticsm of his formulation of entropy by Ernst Zermelo (1871–1953) particularly irksome [156]. It was based on Henri Poincaré's (1854–1912) theorem, which states that all mechanical systems must eventually return to points in $(\mathbf{r}, \mathbf{v})$-space arbitrarily close to the initial point (see [22, pp. 9, 443] and [294]). Boltzmann's friend Josef Loschmidt (1821–1895) also pointed out that the reversibility of Newton's laws fundamentally denies the possibility of defining an irreversible function based on those laws [22, pp. 9, 443].

Boltzmann's final formulation of this theorem used a statistical representation of $H(t)$ [18–20]. To illustrate Boltzmann's statistical concept for the time dependence of $H(t)$, Fig. 1.6 shows results for $H(t)$ produced in a Monte Carlo simulation of a gas relaxing from a state in which all the atoms had the same kinetic energy to a Maxwellian distribution. We notice the rapid and random variations in $H(t)$ on a short time scale, even after equilibrium has been reached.

Boltzmann's ideas on the meaning of $H(t)$ and his identification of the entropy as [48, p. 78]

$$S = -kH(t \to \infty) \tag{1.16}$$

developed over a period of time (see [18–22]). An extensive treatment of his ideas appeared in his 1877 paper, which contains some rather detailed discussions of his understanding of entropy for gases in equilibrium. In the first of these discussions (Section I of the paper) Boltzmann used discrete units of energy, which he considered

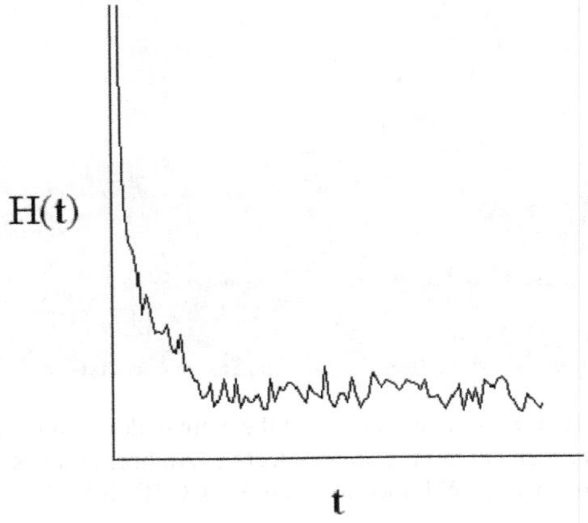

Fig. 1.6 Plot of Boltzmann's function $H(t)$ relaxing to equilibrium. Data from a Monte Carlo simulation of a gas relaxing from a state in which all atoms have the same kinetic energy to a Maxwellian distribution. [Calculations on Maple 10]

unphysical, although mathematically convenient. Then in Section II he carried out the calculation for continuous energy, obtaining the same result.[15]

Boltzmann's formulation of the H-theorem and his statistical interpretation of the behavior of discrete atoms was a great step forward for theoretical physics. David Mintzer (b.1926) points out that the smooth behavior of $H(t)$ represents the ensemble averaged behavior. A single system in the ensemble will have a single-partcle distribution function that deviates from the ensemble averaged distribution (see Sect. 1.5). The H-function for a single system will then deviate in a rapid, time-dependent fashion around the value of the H-function for the ensemble. Our calculation resulting in Fig. 1.6, based on a Monte Carlo simulation, confirms the particulate nature of the gas, as did Boltzmann's 1877 study.

1.4 Gas Discharges and Cathode Rays

The flow of electric current in an electrolyte was understandable in terms of chemistry. The problem of the flow of current in rarefied gases, however, proved to be more complicated.

William Watson (1715–1787) was the first to record an observation of electrical current in a rarefied gas. He placed the poles of an electrostatic generator in an evacuated tube and in a darkened room observed a marvelous light display, which reminded him of the *aurora borealis*. He interpreted what he was seeing as pure electricity flowing in the tube and published these observations in the *Philosophical Transactions of the Royal Society* in 1748 [289, p. 390].

Almost a century later, in 1838, Faraday conducted a similar experiment in rarefied air. He observed what appeared to be a purple haze proceeding from the positive pole

[15] The translators Sharp and Matschinsky refer to this as "hedging his bets."

Fig. 1.7 Southern facade of the main building of the University of Bonn in 2007. This was formerly the residence of the prince-elector. [Author: Thomas Wolf (Der Wolf im Wald). Thomas Wolf, www.foto-tw.de/ Wikimedia Commons/CC BY-SA 3.0]

and stopping just short of the negative pole. A dark space appeared between the purple haze and the negative pole (cathode). This dark space was subsequently referred to as *Faraday's dark space*.

1.4.1 Rarefied Gases

Research on currents in rarefied gases stalled because of the lack of good vacuum pumps. Then in 1855 Heinrich Geissler (1814–1879) invented the mercurial air pump [236]. Geissler's pump allowed an investigator to produce extremely low pressures in a glass tube with electrodes at each end of the tube and study the flow of electric current through the tube under various degrees of rarefaction of the gas [289, p. 392].

Geissler pumps were, however, expensive and difficult to use and he had difficulty disseminating them until he began a collaboration with Julius Plücker (1801–1868), a professor of mathematics and physics at the University of Bonn. Plücker's interest was in the dynamics of chemical reactions, which he chose to study spectroscopically. Figure 1.7 shows a photograph of the southern facade of the main building of the University of Bonn taken in 2007.

It became increasingly clear that he needed a collaborator with sophisticated skills in chemistry and experimental technology, and in the early 1860s Plücker invited his former student Johann Wilhelm Hittorf (1824–1914) to join him in his spectroscopic studies. In 1847 Hittorf had accepted a *privatdozent* position at Münster where he later became professor of physics and chemistry and began a research program in electrochemistry [203, pp. 214–224].

Hittorf thought that studies of conductivity (Ohm's law) would reveal the character of the effective phenomena.[16] Maxwell's kinetic treatment of gases had considered essentially all gas phenomena except electrical conductivity[17] [203, p. 232].

1.4.2 Cathode Rays

In 1869 Hittorf found that, if he placed a solid body in the tube between the cathode and the glowing gas, a shadow was produced. He correctly concluded that the glow was produced by rays emerging from the cathode, which caused a phosphorescence when they struck the tube walls. These were thus called *cathode rays* [289, pp. 232, 393].

In 1876 Eugen Goldstein (1850–1930) of the Berlin Observatory found that cathode rays carried energy and were emitted perpendicularly to the cathode surface. By perforating the surface of the cathode with holes or channels (*Kanalen*), Goldstein also discovered that fluorescent spots appeared on the glass tube behind the cathode. These he termed *Kanalstrahlen* or canal rays [124].

The English engineer Cromwell Varley (1828–1883) had already suggested that the cathode rays were particles of matter comprising a torrent of electricity projected from the cathode [277]. The idea of a torrent of negative charges was also advocated by William Crookes (1832–1919). In his investigations of rarefied gas discharges Crookes was able to identify a dark space (Crookes' dark space) between the cathode and a glow that separated this dark space from Faraday's dark space. He correctly deduced that there were no collisions of the cathode rays with molecules in this first dark space and that the glow was a result of such collisions. However, he thought that the cathode rays were ordinary molecules carrying charges. Because the mean free path of the cathode rays was so long in Crookes' dark space, he considered them to form a fourth state of matter.

At the same time, certain results raised questions about the ideas of Varley and Crookes. The beams did not seem to deflect under the influence of electric and magnetic fields, nor did they produce measurable electric and magnetic fields, as currents would be expected to do, and they were able to pass though thin metal films, something particles would not be expected to do. George F. FitzGerald (1851–1901) cautioned against simple reasoning about the supposed torrents of particles, pointing out that the space surrounding such torrents might contain other electrical effects that would screen them.

---

[16] Heinrich Hertz was one of the few who supported Hittorf's approach. Most scientists were skeptical of the role of gases in conductivity and indeed some thought the vacuum to be an ideal conductor, while gases served only as insulators [203, p. 232].

[17] Efforts with ionized gases (plasmas) in the twentieth century led to an understanding of electrical conductivities and energy transport in these systems. A missing component in the investigation of the ionized gas in the nineteenth century was the electron.

Fig. 1.8 Entrance to the Cavendish Laboratory at the original laboratory site on Free School Lane. [Author: William M. Connolley. Permission under GNU Free Documentation License.]

In 1871 the Cambridge University Cavendish Laboratory was established under the direct supervision of Maxwell [36]. Figure 1.8 is a recent photograph (2005) of the entrance to the original Cavendish Laboratory site on Free School Lane.[18]

The laboratory won recognition as one of the world's great laboratories for experimental physics under Joseph J. Thomson[19] (1856–1940), who was appointed Cavendish Professor of Physics at Cambridge University and director of the Cavendish Laboratory in 1884.

Thomson was not himself a particularly good experimentalist. His earlier work had been primarily in mathematical physics. However, he had a good sense of what constituted important experiments and how to design and perform them. He was also a good leader of research teams [143, p. 385].

Thomson supported the idea of a particle nature for cathode rays, noting that a charged particle entering one face of a thin metal film could cause a second charged particle of the same character to emerge from the other side [289, p. 396]. However, at that time our understanding of electricity and magnetism was dominated by Maxwell's field theory (see [181]), which gave a general sense that electric and magnetic phenomena were of a continuous nature. This stood at odds with the particulate pictures of Varley, Crookes, and now Thomson.

---

[18] Free School Lane is a historic street in Cambridge, England, where several important university buildings are located. Among them is the physics department's Cavendish Laboratory. The name of the street comes from the "Free School" which was established in the seventeenth century by Dr Stephen Perse who left money in his will to educate 100 boys from Cambridge, Barnwell, Chesterton, and Trumpington.

[19] At the end of the nineteenth century there were two famous English physicists with the family name Thomson: J.J. Thomson and William Thomson (Lord Kelvin). We will include first names or initials only when there is a possibility of confusion.

Von Helmholtz was cautious, however. In a lecture he delivered to the Chemical Society of London in 1881, he pointed out that a particulate concept of electricity was entirely consistent with the ideas of put forward by Faraday on which Maxwell's theory was ultimately based. Specifically, he pointed out that, if one accepted an atomic picture of matter, as the majority of chemists did in 1881, then to be consistent one must accept that electricity is composed of both positive and negative parts, which would thus be particulate in nature [289, p. 397].

In 1884 Arthur Schuster (1851–1934) presented a general theoretical picture of current flow in gases, which was the result of dissociation of gas molecules [255]. This led Schuster to support a particle picture for cathode rays. The picture was beginning to emerge that liquids and gases were not fundamentally different in the way they conducted electrical current.

By that time, research into the conduction of electricity in gases had become a central topic in experimental physics. There was great promise that the fundamental properties of matter might be discovered in this way. And in 1893 William Thomson (Lord Kelvin) asserted that experiments on electricity in high vacuum might well yield the first step in understanding the relationship between aether and ponderable matter [272]. Then in 1894, J.J. Thomson measured the velocity of cathode rays using a rotating mirror technique, finding it to be much less than the velocity of light. This meant that cathode rays could not be vibrations in the aether [266].

In a lecture at the Royal Institution in April of 1897, J.J. Thomson attempted to reconcile the particle concept of cathode rays with Philipp Lenard's (1862–1947) experiments on the passage of cathode rays through matter [170, 267]. Lenard's measurements suggested that cathode rays had a mean free path in air of 0.5 cm. The mean free path of a molecule in air was known to be of the order of 10^{-5} cm. Taken together with the response of cathode rays to electric and magnetic fields in high vacuum, Thomson concluded that cathode rays were charged particles that were very much smaller than atoms. This implied that the atom might itself be composed of smaller parts, an idea Thomson and others found quite startling. It motivated FitzGerald to suggest that cathode rays might actually consist of free electrons. The idea of an electron, and the name, originated with George J. Stoney (1826–1911), who proposed this particle at the 1874 meeting of the British Association for the Advancement of Science in Belfast, based on an analysis of chemical reactions [260].

1.4.3 Thomson's Experiment

The debate around this issue was brought to an end by an experiment conducted by J.J. Thomson in 1897. The apparatus he designed for the experiment is sketched in Fig. 1.9.

This apparatus was basically a glass tube containing a gas under high vacuum. The anode collimator was made from two discs, each with a small hole in the center, and produced a narrow cathode ray beam which struck the bulb at the far end of

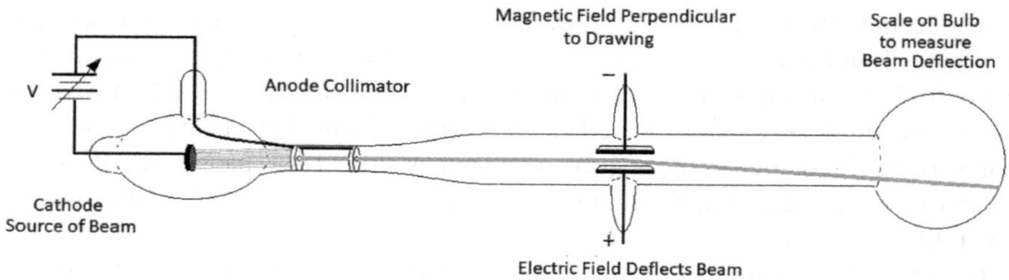

Fig. 1.9 The experimental apparatus used by J.J. Thomson to measure the charge-to-mass ratio of the electron. [Drawn by CSH]

the tube. Thomson proposed to show that the cathode rays were actually Stoney's electrons. Their velocity was determined by the setting on the voltage V between the cathode and the anode. Thomson had a scale on the bulb to indicate the deflection of the cathode ray beam. In a small region toward the center of the tube, two flat disks produced a uniform electric field which deflected the beam. Two coils (not shown) produced a uniform magnetic field perpendicular to the electric field. The current in the coils determined the magnitude of the magnetic field. Thomson could then easily adjust the magnitude of the electric and magnetic fields from outside the tube.

For a fixed value of V, Thomson conducted the experiment in two steps. He first obtained the velocity of the (supposed) electrons in the beam from the ratio of the electric and magnetic fields required to produce no net deflection of the beam. He then turned off the magnetic field and measured the deflection of the beam due to the force from the electric field. In this case, the electrons travelled in a parabolic path when moving through the region between the plates. The remainder of the path, through the region where there was no force, was thus a straight line. From the beam deflection he could determine the time of transit between the plates. From these data he was able to obtain the ratio of the mass-to-charge (m/e) of the electron[20] [268].

Thomson's design of this apparatus and all his calculations were based on his assumption that the cathode rays were actually a torrent of electrified corpuscles, as he termed them. His treatment of the cathode rays as particles formed the basis for the measurements he made and for his interpretation of those measurements. The cathode rays were clearly behaving exactly as one would expect for electrified corpuscles. It would have been very difficult to insist that they were anything else.

In 1906 Thomson was awarded the Nobel Prize in Physics "in recognition of the great merits of his theoretical and experimental investigations on the conduction of electricity by gases."

The work of Goldstein showed that the cathode rays were not the only things present in a gas discharge. There were also canal rays. Wilhelm Wien (1864–1928) began an extensive study of these canal rays. He found that they did not respond

[20] Thomson had actually calculated the ratio m/e for the lecture at the Royal Institution in April. The present experiment used different gases and was more accurate, although the differences in the results were only in the second decimal place.

as readily as the cathode rays to magnetic fields, something he attributed to their particulate components having a considerably higher mass-to-charge ratio or a much greater velocity than the cathode rays. Through collisions of the canal rays with gas molecules he found that these rays could have positive, negative, or neutral charges [285, 286]. It was only a short step from these experimental results to the conclusion that a sum of the canal rays and the electrified corpuscles would result somehow from the gas initially present.

1.5 Statistical Mechanics

1.5.1 Gibbs' Perspective

At the dawn of the twentieth century, the American recluse and first professor of mathematical physics at Yale College, Josiah Willard Gibbs (1839–1903), summarized the last years of his life's work on a topic he called *statistical mechanics* in his classic monograph *Elementary Principles in Statistical Mechanics* published in 1902 [123]. In the preface to that monograph he wrote [123, p. viii]:

> The laws of thermodynamics, as empirically determined, express the approximate and probable behavior of systems of a great number of particles, or, more precisely, they express the laws of mechanics for such systems as they appear to beings who have not the fineness of perception to enable them to appreciate quantities of the order of magnitude of those which relate to single particles, and who cannot repeat their experiments often enough to obtain any but the most probable results.

He noted, also in the preface, that the specific heat predictions of theory were not matched by measurements [138, pp. 177–178]. He therefore wrote:

> Difficulties of this kind have deterred the author from attempting to explain the mysteries of nature.

He thus contented himself with a more modest study, restricted to the statistical branch of mechanics [123, p. x].

1.5.2 Ensembles of Systems

Rather than speaking of probability density functions for atoms possessing certain velocities, averaging over these density functions, and multiplying by the number of atoms to obtain system properties, Gibbs considered collections of systems with identical macroscopic (measurable) properties. He called the collection of all systems with the same macroscopic properties the ensemble. Each system in the ensemble was represented by a point in phase space, which had one axis for each canonical coordinate and momentum for each atom in the system. He denoted the density of the representative points in phase space by D and the number of systems by N, then defined $P(\Omega) = D/N$, where Ω is a point in phase space, and

$$1 = \int \mathrm{d}\Omega \, P(\Omega) \ . \tag{1.17}$$

The ensemble average of a system property was the sum over the values of this function for all systems divided by the number of systems in the ensemble. Hence, if the value of the property Φ at the phase space point Ω is $\Phi(\Omega)$, then the ensemble average of the property Φ is

$$\langle \Phi(t) \rangle = \int_{\Omega} \mathrm{d}\Omega \, \Phi(\Omega) \, P(\Omega) \ , \tag{1.18}$$

where the brackets $\langle \cdots \rangle$ designate an ensemble average, i.e., an average over all systems of the ensemble. This Gibbs identified as the thermodynamic (macroscopic) property of the ensemble of systems. We have included the time t in the final result to indicate that we can in principle include nonequilibrium systems [123, p. x].

An example that will be of importance to us is the statistical mechanical formulation of the entropy, which is

$$S = -k \langle \ln P \rangle \ . \tag{1.19}$$

We note that the definition (1.18) is not a probabilistic statement. It is a statistical statement. This is the principal difference between statistical mechanics and kinetic theory. Kinetic theory is a probabilistic approach to the behavior of the atoms of a system. Gibbs' statistical mechanics encompasses kinetic theory.

The properties of a single system in the ensemble differ from the properties of the ensemble. The difference between a system property F and the ensemble averaged property $\langle F \rangle$ is viewed as a fluctuation δF in the property and is defined as

$$\delta F = F - \langle F \rangle \ . \tag{1.20}$$

Although the ensemble average of the fluctuation $\langle \delta F \rangle$ vanishes, the ensemble average of a product of fluctuations $\langle \delta F \delta G \rangle$ does not vanish.

Gibbs considered only equilibrium systems because it was, and still is, beyond our capability to solve the equations of motion for an entire system [123, Chaps. I–IV]. In a deceptively thin volume, Gibbs laid out this beautiful theory which is still the basis of our treatment of microscopic systems. The limits he accepted meant that he only proposed a general Hamiltonian mechanics without specifying the exact form of the Hamiltonian. For this reason Gibbs' statistical mechanics survived the quantum revolution unscathed.

1.5.3 *Statistics of the H-Theorem*

In our discussion of Boltzmann's understanding of the H-theorem, we cited Mintzer's clarification of the rapid positive and negative variations of $H(t)$. He had realized that this results from the difference between the theorem written for the ensemble average and for the single system in the ensemble. Written in terms of the ensemble averaged form of the single particle distribution function, which solves the Boltzmann equation (1.13), the function H is the (negative of) thermodynamic entropy. If we write H in (1.15) for a single system in the ensemble, with (1.20), we have

$$H(t) = \int \mathrm{d}\mathbf{v}\, \{[\langle f(t)\rangle + \delta f(t)] \ln [\langle f(t)\rangle + \delta f(t)]\}$$
$$\approx \langle H(t)\rangle + \delta H(t)\ , \tag{1.21}$$

where $\langle H(t)\rangle = \int \mathrm{d}\mathbf{v}\, \langle f\rangle \ln \langle f\rangle$ and

$$\delta H(t) = \int \mathrm{d}\mathbf{v}\, [\langle f(t)\rangle + \ln \langle f(t)\rangle]\, \delta f(t)\, . \tag{1.22}$$

The function $\delta H(t)$ may be either positive or negative. The function $H(t)$ in (1.21) then expresses the statistical variations of Boltzmann's kinetic theory formulation of entropy in terms of Gibb's statistical mechanics. Even at equilibrium, $H(t)$ varies, because no system behaves exactly like the ensemble average.

1.6 Thomson's Atomic Model

There were significant attempts to build models of the atom in the eighteenth and nineteenth centuries. We have already considered Boscovich's model of 1758, which introduced fields around point particles to form the atom.

In 1815 William Prout (1785–1850) noticed that atomic weights seemed to be whole number multiples of the atomic weight of hydrogen. This led him to consider what he called the *protyle*,[21] which was basically a hydrogenic building block for atoms. Prout's initial observation for low atomic weights did not hold for higher atomic weights. But his proposal resulted in some very accurate determinations of atomic weights, which were of a great benefit to chemistry (see [231] and [143, p. 296]).

In a series of novel experiments, the American physicist Alfred M. Mayer (1836–1897) showed that magnetic needles with their like poles inserted in corks and floated in water arranged themselves in ordered structures under the influence of a central

[21] At its Cardiff meeting in 1920, the British Association for the Advancement of Science accepted Rutherford's suggestion that the hydrogen nucleus be named the "proton," following Prout's word *protyle*.

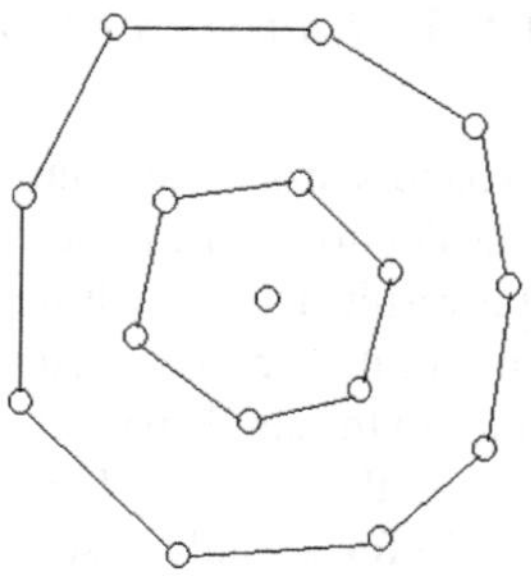

Fig. 1.10 Pattern from a Mayer experiment with 16 magnetic needles (open circles). Slight asymmetry and connecting lines in the original [183]. [Drawn by CSH]

magnetic field. Because of the way the like poles of the needles were arranged, the force between each pair of corks was repulsive. Figure 1.10 shows a representation of 16 magnetic needles as Mayer observed them in 1878 [183]. The asymmetries in Fig. 1.10 appear in Mayer's paper.

These experiments impressed Thomson,[22] who was reflecting on the problems associated with modeling atoms made up of electrified corpuscles [183].

Thomson had already been thinking about a model of the atom before he carried out his experiment to find the mass-to-charge ratio of the electrified corpuscles constituting the cathode rays. Considering the amount of creative energy he put into the development of an atomic model based on those corpuscles, we could almost argue that the Nobel Prize experiment was incidental to his real objective, which was the formulation of an atomic model. Thomson devoted three and a half pages of his *Philosophical Magazine* paper of 1897 to speculative discussions about the possible role of his electrified corpuscles in the structure of atoms. In those pages he discussed Prout's ideas and devoted space to Mayer's results as possible guides to stabilizing the electrified corpuscles [268].

John L. Heilbron (b. 1934) points out that Thomson's efforts to construct a detailed atomic model were a natural part of English physics at this time. English pedagogy in physics simply considered a theory to be incomplete without an accompanying model or some analogy worked out in detail. There had to be something that could be fully grasped. This was very clearly stated by FitzGerald, who wrote: "A Briton wants emotion in his science, something to raise enthusiasm, something with human interest." The model conceived by J.J. Thomson was in the spirit of William Thomson (Lord Kelvin), who had repeatedly insisted that his models were "a disjointed series of tableaux which appeal to the imagination." [129, pp. 41–45]. This approach had the advantage of coming up with a concrete idea that could be tested experimentally.

In 1904 Thomson published a detailed description of a model based on the electrified corpuscles he had found in 1897 and which he believed were most probably the building blocks of the atom [269]. This is the model which is, unfortunately, often derisively referred to as the "plum pudding" model. There is no hard evidence that Thomson himself ever used that terminology. Even a casual survey of the 28 page paper, which is devoted primarily to mathematical arguments, should be sufficient

[22] In this section we are discussing the work of J.J. Thomson, not William Thomson (Lord Kelvin).

to convince us that this paper is not the result of casual speculation. With this publication, Thomson gave the physics community a serious and detailed account of how the atom might be constituted [269].

In the publication Thomson referred to the charged particle he had discovered as a corpuscle, even though Stoney had already introduced the term "electron," which later became the generally accepted term. To avoid confusion we will use the term "electron" for this corpuscle, following Helge Kragh (b. 1944) in his extensive discussion of Thomson's paper [159]. We will also use the modern term "charge" rather than Thomson's term "electrification."

Thomson assumed that the mass of the atom was the sum of the masses of the electrons present in the atom. To balance the electrical charge of the electrons he chose to place them within a uniform sphere made from some positively charged medium which was itself massless.[23] This medium was only there to balance the charge of the electrons and to provide a stable binding force on the electrons. The mass of the atom was at least (in the case of hydrogen) 1836 times the mass of the electron. So many thousands of electrons were required to provide the atomic mass in Thomson's model. This large number of electrons was also supposed to provide the basis for the complicated spectra of different atoms, with their thousands of emission and absorption lines.

The details of Thomson's model were based on his speculations in the 1897 paper. The electron took on the role of Prout's protyle and Thomson was inspired by Mayer's floating magnets to seek a stable configuration of the electrons within the spherical distribution of positive charge. The electron orbits were chosen to be distributed symmetrically as rings in planes perpendicular to a central axis of the atom. Thomson found that stable configurations for more than 5 electrons in a ring could only be found if there were also electrons inside the ring. The numbers of electrons required to be inside the ring increased with the number on the ring itself. For more than 15 ring electrons, the number inside the ring had to be greater than the number actually on it. For example, a stable ring of 40 electrons required 232 inner electrons. He denoted ring electrons by n and inner electrons by p.

Thomson treated the forces among electrons and between electrons and the positive sphere to be Coulomb. The rings of electrons rotated about the central axis formed by a diameter of the sphere of positive charge. For rings containing 1–7 electrons arranged uniformly around the central axis, he obtained the equations of motion of the electrons. Then he considered small perturbations of the electrons around their positions on each ring to establish equilibrium conditions. For what would have been a more general (and more realistic) situation, he accepted that the rings might lie in planes rotating about several different such axes. But at the time of publication he stated that he had not been able to carry out the corresponding calculations [269, p. 255].

For the general case of N electrons in an atom, Thomson considered that a ring containing n electrons required $p = f(n)$ electrons within the ring for stability.

[23] Today we acknowledge electric charge to be a property of a particle. This was not the case in 1904.

If the outer ring had n_1 electrons, then p electrons had to be within the ring and $N = n_1 + f(n_1)$ could be solved for the number n_1 that would make the ring stable. When the solution to this equation was not a whole number, Thomson suggested using the integer part of the solution. The second ring would have some number n_2 of electrons and the number of electrons inside the second ring would be $p_2 = f(n_2)$. Then $N - n_1 = n_2 + f(n_2)$. The idea was once again to use the integer part of the solution. In this way concentric rings could be built up for each plane. For example, for the set of rings with $n = 35$, one obtains $n_1 = 16, n_2 = 12, n_3 = 6, n_4 = 1$.

Having formulated the kind of calculations that could deliver the internal structure of an atom, Thomson went to some lengths to describe how some of the known properties of atoms might be accounted for in his model. He was particularly interested in the periodicities revealed by the Periodic Table, ionization, and electron capture. Because the differences between the atomic masses of different atoms were so much greater than the mass of the electron, he met with little success in describing differences between atoms like helium and lithium. And because his calculations were limited, his attempts to explain atomic periodicities cannot be considered successful either. However, his attempts to explain electron transfer in the formation of ionic compounds gave better results [269].

Although Thomson began his analysis with rather careful consideration of the question of electromechanical stability, he nevertheless required a certain leeway here in his attempts to explain known phenomena. His approach sat rather uneasily with the laws of classical physics in this respect. We will encounter this same issue when we come to consider Niels Bohr (1885–1962) and his atomic model. The difference is that Bohr's approach went on to accept a violation of classical electrodynamics, while Thomson did everything to avoid going that far [15, 269].

Thomson made no direct attempt to describe electromagnetic emission from an atom based on his model, except in a final remark in the paper regarding radioactive materials. He recognized implicitly that there would be energy loss from the rotating rings of electrons. But he seemed to accept that for large numbers of electrons in the rings, the rate of this loss would be extremely small. He then imagined that, at a certain point, the energy loss might require the atom to shed a whole ring of electrons and thereby become a new atom. A ring in an atom of high atomic number might contain enough electrons to make up an alpha particle, and it could certainly yield beta particles, which were just electrons [269].

Thomson's thoughts on the structure of the atom were widely known in the physics community.

1.7 Nagaoka's Atomic Model

Apparently inspired by Maxwell's analysis of the rings of Saturn, Hantaro Nagaoka (1865–1950) produced a Saturnian model of the atom in 1904. Nagaoka's model had the positive charge concentrated in the center of the atom and the electrons circling around it. He rejected Thomson's concept of a positive background charge and sought to explain spectral emission on the basis of perturbations in the electron orbits. His model, however, was mechanically unstable: small perturbations would grow until the system broke up [204].

Nagaoka had studied under the English physicist Cargill Knott (1856–1922) at Tokyo University, continuing to work with him on magnetism after graduating in 1887. This meant that he was not unfamiliar with the English approach to model building. And then in 1893 Nagaoka spent some time at the universities of Berlin, Munich, and Vienna, during which he visited the Cavendish Laboratory. On this extended trip he also attended the First International Congress of Physicists in Paris in 1900, where he was inspired by a lecture by Marie Curie (1867–1934) on radioactivity. He returned to Japan in 1901, taking up an appointment at Tokyo University, where he remained until 1925 (see [129, pp. 52–53] and [214]).

1.8 X-Rays and Electron Number

In 1895, while pursuing studies on gas discharge tubes, Wilhelm Röntgen (1845–1923) serendipitously discovered X-rays. Perhaps because of their widespread applications, within a year of the discovery, more was known about X-rays than about cathode rays [289, p. 404].

1.8.1 Barkla

Charles Barkla (1877–1944), a student of Thomson's, had the idea that information on the structure of atoms could be gathered by probing them with Röntgen's X-rays. Barkla carried out these experiments with Charles A. Sadler (1882–1920) over the period of 1908–1909 [5]. He used Thomson's model to attempt to understand the scattered radiation from atoms.

Part of the scattered radiation was characteristic of the chemical element being studied, and it was polarized. For lighter elements the direction of emission of this characteristic radiation was predominantly in a plane perpendicular to the incoming beam. This was what Barkla expected if the radiation were a result of the acceleration of electrons in the Thomson model (see [139, pp. 327–331]). Barkla noted that this directionality of the radiation was lost when the atomic number of the target atom began to increase. However, this discovery regarding the secondary X-radiation from atoms provided no problem for Thomson's model [9].

The discrepancy between Thomson's model and Barkla's experiments was due to absorption from the primary X-ray beam. Barkla used a standard linear absorption model and found that the absorption coefficient for the primary beam was proportional to about twice the atomic number (see [5] and [129, pp. 36–37]). This was disastrous for Thomson's model, which required many more electrons than the atomic number [9, 10].

1.8.2 Diffraction

The discovery and development of X-ray diffraction compounded this difficulty with Thomson's model. In 1912 Max von Laue (without the "von" at that time) (1879–1960) was *Privatdozent*[24] at the Institute for Theoretical Physics of the Ludwig Maximilians University in Munich, under the direction of Arnold Sommerfeld (1868–1951). Laue had long been interested in the passage of electromagnetic wave energy through conducting media. [see [168]] It was while walking with Paul Ewald (1888–1985) from the institute through the *Englischer Garten* toward Laue's home, that he became particularly interested in the possibility of scattering X-rays in crystals.

Ewald had contacted Laue to raise some questions about his dissertation, which dealt with the dispersion and refraction of electromagnetic waves by electrons in a crystal lattice. On the walk, Laue pressed Ewald on the spacings between the atoms in the crystal he was using. Ewald did not know exactly, except that they were very small. Laue was thinking of X-rays, but Ewald said later that he could get nothing out of Laue regarding the reason for his interest. [291] Then on 12 April 1912, Laue obtained a diffraction pattern by directing an X-ray beam at a crystal. [91] The experiment initiated by Laue was conducted by his assistants Walter Friedrich (1883–1968) and Paul Knipping (1883–1935) [109, 119].

The X-ray beam was passed directly through the crystal and recorded on a photographic film placed perpendicular to the beam. The developed film revealed the central beam and a regular array of spots around it. Today such an array is known as a *Laue pattern* [58, pp. 24–26]. A one-page note announcing the results of the experiment was deposited by Sommerfeld with the Bavarian Academy of Science on 4 May 1912, before the more comprehensive paper appeared[25] [91].

1.8.3 The Braggs

While walking beside the river and thinking of the results of Friedrich and Knipping's experiments, William Lawrence Bragg (1890–1971), a first year research student at Cambridge University, realized that the spots could result from reflections of the central beam by the crystal planes. He checked his idea by performing calculations based on the spots that appeared in Friedrich and Knipping's photographs [32].

William Henry Bragg (1862–1942), the father of William Lawrence, reasoned that the cleavage planes of a crystal would lie along planes of symmetry. When the angle between the X-ray beam and the crystal plane was equal to that between the detector and the plane, as required for reflection, he found reflective peaks at distinct angles.

[24] The German university system does not easily translate into the American one. A *Privatdozent* is someone who has written a *habilitation* and is granted the right to teach and lecture. The Privatdozent may be thought of as someone with the qualifications of an associate professor, although often without a faculty appointment.

[25] Laue received the 1914 Nobel Prize "for his discovery of the diffraction of X-rays by crystals."

This not only corroborated his son's ideas, but also showed that the X-rays were behaving as waves. This established a new scientific methodology: X-ray diffraction for the measurement of lattice spacings in crystals.[26]

The Laue and the Bragg approaches were fundamentally different, although the experimental results provided the same information. Laue's approach was electrodynamical, treating the electromagnetic radiation as scattered from the electrons in the crystal. According to Thomson's atomic model, those electrons were in the atoms that made up the crystal.

Due to the periodicity of the lattice, it was natural to introduce a Fourier transform to represent the density of scattering centers (electrons). In the Bragg approach, the lattice planes were identified using *Miller indices*, which are reciprocals of the spatial variables in the characteristic geometry of the crystal. The Miller indices of a lattice plane then corresponded to a direction in the Fourier wave vector space. As Ewald noted, it was difficult to make the connection at first, but once the method had been found it became rather simple [291]. The result was a demonstration of the equivalence of the Laue and Bragg approaches to X-ray diffraction.[27]

For our purposes here, it is only important to realize that the Laue approach provides a measure of the number of electrons in the scattering centers, that is, in the atoms. Laue's experiments then provided a crucial check on the number of electrons in an atom. The results were much more conclusive than the measurements of linear absorption coefficients made by Barkla. The fate of Thomson's model was thus finally sealed by X-ray diffraction. The large number of electrons that characterized Thomson's model were quite simply impossible according to the X-ray diffraction results.

1.9 Rutherford and the Nucleus

Ernest Rutherford (1871–1937) was from the colonies, in fact, from New Zealand. This would make a difference for him in England, even though he was a brilliant student, passing the Tripos examination in mathematics with first class honors, and thus becoming what is known as a wrangler.[28] Kelvin and Maxwell were wranglers. But that alone would not be enough to net Rutherford a university position in England.

So after graduate work at the Cavendish Laboratory, Rutherford took up a post at McGill University in Montreal, Canada. At McGill he carried out important work on nuclear transmutation, and this eventually gained him a position at the University of Manchester in the industrial city of Manchester, England, in 1907. There he led

---

[26] The Braggs received the 1915 Nobel Prize "for their services in the analysis of crystal structure by means of X-rays" [91].

[27] The book by Harald Ibach and Hans Lüth [150] contains a particularly nice explanation of the connection between the Laue and Bragg theories.

[28] Heilbron identifies a *wrangler* as someone who can survive, with first class honors, a stiff week-long examination in mathematics held in an unheated room in January [129].

a highly productive research laboratory. In 1908 he was awarded the Nobel Prize in Chemistry [212] "for his investigations into the disintegration of the elements, and the chemistry of radioactive substances."

In Manchester, Rutherford began studying the scattering of particles from thin films of heavy metals. His assistant in this work was Hans Geiger (1882–1945), the John Harling Fellow at the University. In 1907 it was known that the α-particle was a doubly ionized helium atom. In Thomson's model of the atom, however, that still left some freedom regarding the possible structure of the α-particle. But in any case it was a massive and energetic projectile if the collision target was an electron, and the small scatter (of the order of a degree) or broadening of the collimated beam that Rutherford and Geiger observed was therefore curious. He estimated that this scattering would require an electric field of 100 MV cm^{-1} [238, p. 47].

According to Rutherford, Geiger suggested that they give a "small research project" to Ernest Marsden (1889–1970), who was a Hatfield Scholar under Geiger's care at the time. Geiger specifically suggested that Marsden should look for large angle scatter, although Rutherford personally did not expect Marsden to observe any such thing.

Geiger had developed an electronic counter, but it could only count scattered α-particles within small angular ranges. Geiger and Marsden wanted to look for α-particles scattered over wide ranges so they used the more familiar scintillation technique in which scattered α-particles produce flashes on a zinc sulfide screen. Preparation for gathering data from a scintillation screen required half an hour's adaptation in a totally darkened room [238, pp. 47–48].

According to Rutherford, after two or three days, Geiger came to him in great excitement, saying that they had actually detected back-scatter of the α-particles.[29] Rutherford later remarked that this was the most incredible moment of his life. He said it was as though someone had fired a 15 in shell at a piece of tissue paper and it came back and hit them.

Because no backscatter of α-particles was expected, Geiger and Marsden constructed a very intense α-source by drawing down a conical tube, filling it with radium and closing the end with a mica window. They shielded the scintillation detector from the direct α-beam by a lead plate and placed the target film in a position that allowed only back-scattered α-particles to be detected. The flashes on the scintillation screen were observed with a microscope. Figure 1.11 is a sketch of the apparatus Geiger and Marsden used. Targets were Al, Fe, Cu, Ag, Sn, Pt, Au, and Pb films. Between 3.4 (Al) and 62 (Pb) scintillations per minute were recorded [122].

Using layers of gold foil, because gold is very malleable, they also measured the number of α-particles reflected as a function of the thickness of the foil. This was to determine whether or not the reflection was a surface phenomenon. The data from this experiment are shown in Fig. 1.12.

Reflecting on this result Rutherford soon realized it meant that most of the mass of an atom must be concentrated "in a minute nucleus," which must carry a positive

[29] Marsden recalls the care with which he checked the possible sources of error and his meeting with Rutherford on the steps to Rutherford's "private room"[238, pp. 48–49].

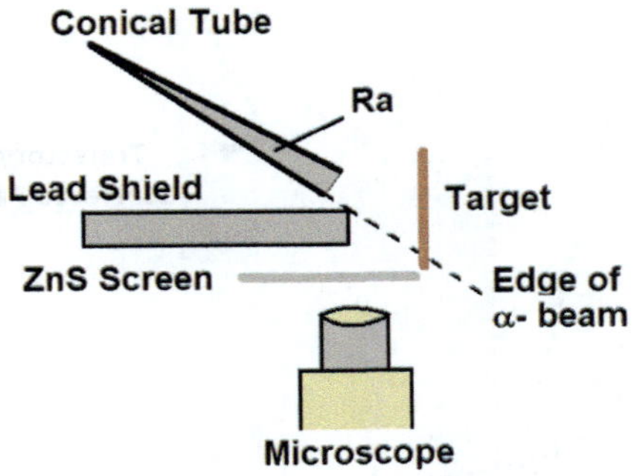

Fig. 1.11 Geiger and Marsden's apparatus (1909). The ZnS scintillation screen was shielded from any α-particles emitted directly from the source (RaA and RaC) and received only α-particles reflected from the metal film target. Scintillations were observed with the microscope. (From Fig. 1 in [122]. Drawn by CSH)

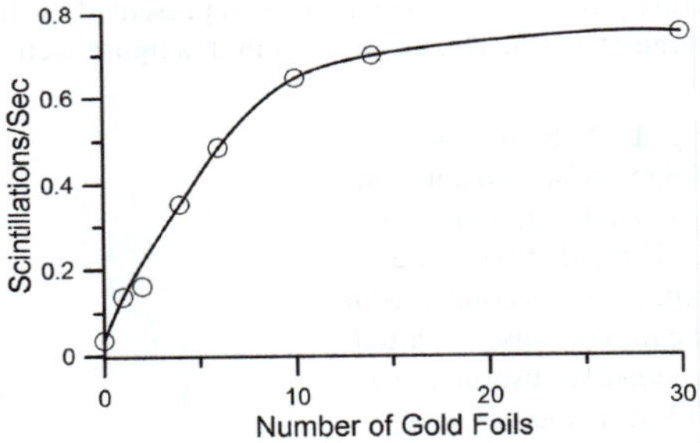

Fig. 1.12 Number of reflected α-particles as a function of the number of layers of gold foil in the target. The data point at 30 was simply from a gold plate. (Data taken from Fig. 2 of [122]. Plotted by CSH.)

charge (quotes from Rutherford in [143, p. 418]). This had to hold for both the target atom and the α-particle, which had already been identified as a doubly ionized helium atom.

Assuming with Rutherford that most of the masses of the α-particle and the stationary target atom were both contained in point-like positive charges, then the trajectories of the α-particles would appear as we have drawn them in Fig. 1.13. The filled circle on the horizontal axis represents the nucleus and the dotted lines represent the trajectories of single α-particles in the collimated beam. Each trajectory has a different impact parameter, which is the distance of the trajectory from the horizontal axis when an infinite distance from the nucleus.

Rutherford worked out the dynamics of this sort of collision. From his analysis he found that the probability that an α-particle would be scattered into an angle Φ from the initial beam direction was proportional to

$$\ell\left(\frac{e_{\mathrm{N}}^{2}}{v_{\alpha}^{4}}\right)\frac{1}{\sin^{4}\frac{\Phi}{2}},$$

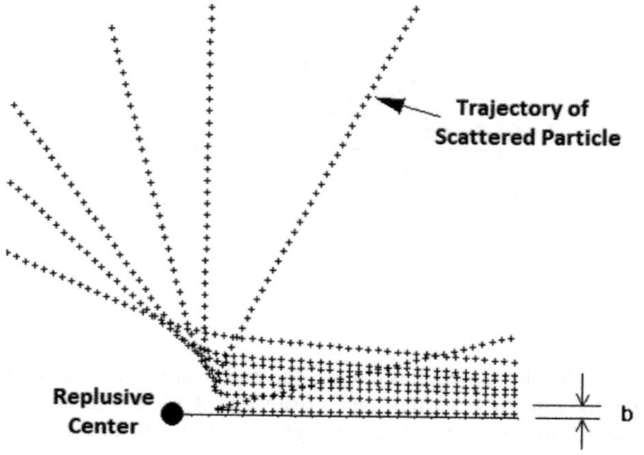

Fig. 1.13 Possible scattering of α-particles in a collimated beam (represented by dotted lines) from a tiny positively charged nucleus (represented by the filled circle on the horizontal axis). The impact parameter is b. The trajectories in this figure were calculated using Maple 10

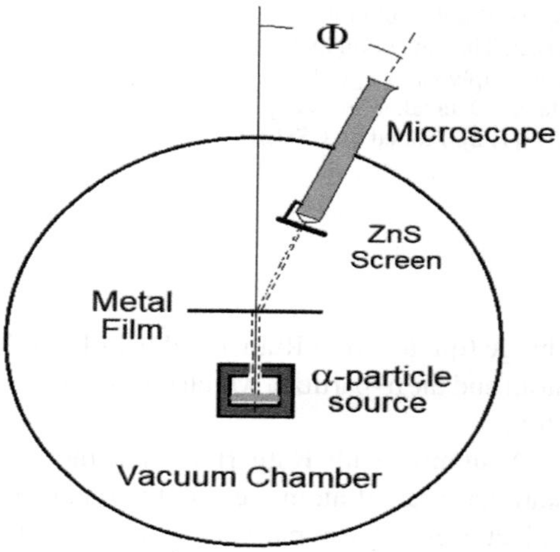

Fig. 1.14 Sketch of the Geiger–Marsden apparatus. The α-particle source is in a lead-shielded box. The experiment is conducted in vacuum because α-particles are readily absorbed in air. This drawing has been slightly enhanced compared with the one appearing in [143, p. 420]. [Drawn by CSH]

where ℓ is the thickness of the target film, e_N is the nuclear charge of the atoms in the target film, and v_α is the velocity of the α-particles in the beam (see, e.g., [140, pp. 106–110]). He later pointed out that this result was verified by Geiger and Marsden in 1913 in a series of "beautiful experiments." (see [122, 241], and [143, pp. 418, 420]). The experimental apparatus used by Geiger and Marsden is drawn in Fig. 1.14.

The results of Geiger and Marsden's experiments and Rutherford's analysis finally did away with the massless positive charge in Thomson's atom.

1.10 Summary

In this chapter we have come a long way, both historically and intellectually. We have been focusing on the atom, suggesting that the idea originated with the Presocratic Greek philosophers, who were themselves physicists. Their primary concern was with what they could see and how they should understand it. As we pointed out, the atom may have been proposed to resolve the question of change. However, there is no easy answer to the question of the origin of this hypothesis. We may ask whether the idea that extremely small, eternal particles could be the origin of all we see would have emerged without Parmenides. But at the end of the day, the question is: what is reality?

At the beginning of the seventeenth century, the arguments for atoms became more sophisticated and were debated by Jesuit mathematicians. Boyle and Newton tried to formulate a picture of atoms that really went nowhere. And then, in the eighteenth century, Bernoulli produced a remarkably simple idea that is close to our present concept of a gas. In the same century, the Jesuit priest Boscovich produced several ideas about the inner workings of atoms that could at least help chemists to understand what they already knew about bonding.

We then turned to the studies of electricity at the end of the eighteenth and beginning of the nineteenth centuries, experimental research programs that were gradually revealing the atomic structure of matter. In a sense these were technological, since they aimed primarily to produce steady voltages. However, the basic concept of the atom was necessary in order to understand the behavior of ionic solutions.

We went on to consider the theories of phlogiston and caloric, which had been developed to explain the phenomena of fire and heat. We chose to look at these in some detail because of their historic importance. It was the demise of these theories that gave us mass conservation in chemical reactions and also thermodynamics. The latter provided a statement of energy conservation and the concept of entropy as a property of systems that increases in any spontaneous process within an isolated system. This set the stage for the development of an atomic kinetic theory. The aim was to establish that an atomic picture of matter obeying classical mechanics could actually yield the laboratory results. In the hands of Ludwig Boltzmann, the kinetic theory produced the beginnings of an understanding of irreversibility in atomic systems.

Geissler's invention of the mercurial air pump made the study of electrical discharges in gases possible. However, these gas discharges could not be understood without introducing the electron as a component of the gas discharge. From there it was almost a logical step to Thomson's experiment with cathode rays, provided Thomson was actually looking for a model of the atom. And we discovered that this was indeed his primary motivation. His discovery that the cathode rays, which were the carriers of electric current in gas discharges, were torrents of charged particles gave him the building block he required for his atomic model.

We followed Thomson's discovery with a discussion of the statistical mechanics of Gibbs. The importance of Gibbs' ideas in any attempt to understand the dynamics of systems at an atomic level cannot be overestimated. We therefore included a brief discussion of this theory at the appropriate historical point in the story. Gibbs' ensemble concept clarifies the difference between macroscopic (thermodynamic) treatments of the properties of matter as a branch of theoretical mechanics and clarifies the meaning of the probabilities in the kinetic theory. Statistical mechanics defines the fluctuation, which was central to the discovery of the quantum.

Recognizing the importance of atomic models to Thomson, as well as to the general physics community at this time, we considered his own model in some detail and then another due to Nagaoka. We also considered the failure of the Thomson model to explain the results of X-ray diffraction. Needless to say, his model failed on other counts, too.

We closed our introduction to atomic theory with the discovery of the atomic nucleus. This transformed the problem of modelling the atom. Our discussion also provided a glimpse of experimental techniques being developed at the beginning of the twentieth century and another look at the interplay between experiment and theory.

Chapter 2
Discovery of the Quantum

I have always looked upon the search for the absolute as the noblest and most worthwhile task of science.

Max Planck

Eleganz sei die Sache der Schuster und Schneider.

Elegance is the concern of cobblers and tailors.

Ludwig Boltzmann

2.1 Introduction

The quantum was discovered by the theoretical physicist Max Planck (1858–1947) at the beginning of the 20th century. Others were certainly involved, including both experimentalists and theoreticians, and their work was crucial. The fact that the problem of blackbody radiation was important for industry would have been a significant motivation, as would Planck's close personal relations with the experimentalists at Germany's main metrological laboratory, the *Physikalisch-Technische Reichsanstalt* (PTR) in Berlin-Charlottenburg.

Planck's understanding of physics and his role as a university professor in an institution considered to be at the heart of German physics are important to understanding his personality. John Heilbron (b. 1934) provides a picture of Planck's sense of his role as a scientist and a German citizen in the 19th and 20th centuries.[1]

Martin J. Klein (1924–2009) presents a detailed reflective description of Planck and the blackbody problem in *Physics Today* [156], an article that may be consulted

[1] See in particular the 2000 edition of his book *Dilemmas of an Upright Man* ([130], pp. 69–70, 205–217), which contains an afterword exploring Planck's response to the tragedy of National Socialism.

C. S. Helrich, *The Quantum Theory—Origins and Ideas*, History of Physics,
https://doi.org/10.1007/978-3-030-79268-8_2

for an overview of German science at the turn of the twentieth century, describing the personalities involved and also Planck's philosophical position. Planck faced an intellectual crisis when he was finally compelled to accept Boltzmann's interpretation of the second law and entropy. Klein gives a more detailed study of Planck's physics and his mathematical development in the *Proceedings of the International School of Physics "Enrico Fermi" Course LVII*.

2.2 Planck and the Second Law

Max Planck held that the highest goal for science was an exact description of nature. His dissertation (Munich 1879) was based on a study of the second law of thermodynamics. For this purpose, he had immersed himself in the thoughts of Clausius, which left a lasting impression on him. As a result he adhered to the thermodynamic approach over the molecular approach ([184], p. 32). We may consider Planck's position on the molecular approach as somewhat curious, since Clausius was one of the original contributors to the kinetic theory of gases (see [40]; [138], p. 127) However, Brush points out that Clausius always tried to keep his work on thermodynamics and kinetic theory well separated. He was concerned that any failures in kinetic theory might harm the validity of thermodynamics ([40], p. 24).

Planck supported Boltzmann in his arguments with energeticists like Wilhelm Ostwald (1853–1932) and Georg Helm (1851–1923). However, he did not support Boltzmann's atomism. And in particular, he did not support Boltzmann's interpretation of the second law of thermodynamics, which he considered to go against the idea of an exact science. Zermelo, whose criticism of Boltzmann's formulation of entropy Boltzmann found particularly irksome, as we recall, was a student of Planck's [156].

The formula for the entropy that appears on Boltzmann's tombstone

$$S = k \log W, \tag{2.1}$$

was first written by Planck [226]. The term W appearing in (2.1) is the number of "complexions" among the molecules of a gas, which is a large number. Although W is understood to stand for *Warscheinlichkeit*, which is German for "probability," a probability is always less than unity, and this would result in a negative value for $\log W$.

2.3 Blackbody Radiation

The process of emission of thermal (electromagnetic) radiation by matter was considered to be one of the great problems in physics at the end of the 19th century. Interest was motivated by industrial requirements for the illumination of cities, more than by scientific curiosity (see [184], pp. 24–33; [188], pp. 19–30).

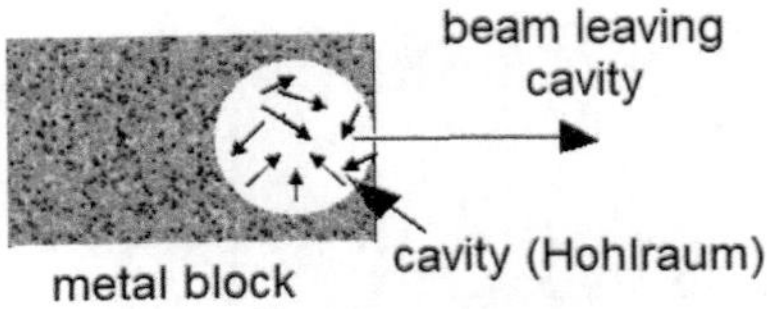

Fig. 2.1 A hohlraum is a cavity within a heated metal block. A small hole is made through which a beam of thermal radiation may escape from the cavity, while very little radiation can enter (Drawn by CSH)

In 1860 Kirchhoff defined the *blackbody* as a physical body that emits its own radiation but reflects no radiation that happens to be incident upon it. Specifically, he provided a proof that, if E is the "radiating power" and A is the "power of absorption" of the body, the ratio E/A is independent of the nature of the body and the underlying emission mechanism. This ratio depends only on the wavelength λ of the radiation and the temperature of the body T. This did, of course, exclude electrically charged bodies and phosphorescent or fluorescent bodies ([154], [155]). Kirchhoff identified the radiation energy density as $u_\lambda = J(\lambda, T)$, noting the great importance of finding this function $J(\lambda, T)$, and pointing out the experimental difficulties involved ([154], cited by Pais [215], pp. 364–365). With Thomson's discovery of the electron as a particle in 1897, it was at least possible to consider the laws governing the motion of a bound electron in matter ([188], p. 30; [184], pp. 35–36).

We can obtain the conditions to produce blackbody radiation in the laboratory by creating a cavity within a solid body, such as a metal block, and heating the body to the temperature at which we wish to conduct the study. The cavity will then naturally be filled with thermal radiation characteristic of the (thermodynamic) temperature of the solid body. If the cavity is made close to the boundary of the solid body we may then bore a small hole through which the thermal radiation can pass with a vanishingly small amount of radiation entering the cavity through the hole. Such an internal cavity is a *holhlraum* (empty cavity, in German). The basic concept of a Hohlraum is sketched in Fig. 2.1. The small arrows inside the cavity represent thermal radiation at equilibrium with the metal. The larger arrow represents the escaping radiation.

In 1889 Planck was named successor to Gustav Kirchhoff's (1824–1887) position at the Friedrich-Wilhelms-Universität in Berlin, and in 1892 became a full professor. Then in 1894 he turned his attention to blackbody radiation because of the effort being devoted to this problem at the PTR[2] ([229]; [39], p. xvii). Planck intended to investigate the problem of the second law in a different way by considering what seemed to be the most fundamental of processes: the interaction of electromagnetic radiation with matter ([157], p. 2).

[2] The *Physikalisch-Technische Reichsanstalt* evolved into the present *Physikalisch-Technische Bundesanstalt* (PTB) after World War 2.

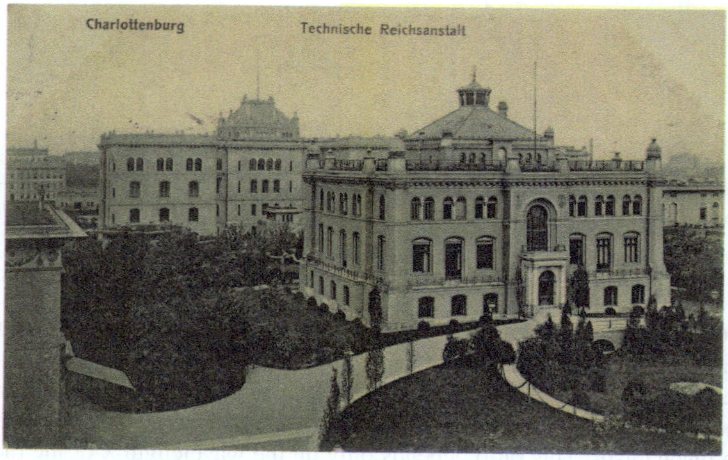

Fig. 2.2 *Physikalisch-Technische Reichsanstalt* in Berlin-Charlottenburg 1913. This was the main industrial/scientific laboratory in Germany at the beginning of the twentieth century

Figure 2.2 shows a postcard photograph of the PTR in Berlin-Charlottenburg taken in 1913. This was the main metrological laboratory in Germany at the beginning of the 20th century. The need for a metrological laboratory was clear after the Metre Convention of 1875, and agreement between the most important industrialized nations of the day. Crown Prince Frederick of Prussia (1794–1863), Werner von Siemens (1816–1892), and Count Moltke (Helmuth von Moltke) (1800–1891) obtained support from the Reichstag (Imperial Diet of the German Empire) for the foundation of the PTR. Its first annual budget was approved in March of 1887. Hermann von Helmholtz, the most famous physicist of his time, was appointed the first president of the new foundation, and took up his duties as director in 1887. The standardization of measurements and the development of instruments was of great importance to a nation basing its future, partially out of necessity, on science [232].

The *National Physical Laboratory* in the UK (1900) and the *National Bureau of Standards* in the US (1901) followed soon afterwards. *Bell Telephone Laboratories* in the US, with roots going back to 1907, was established by an industrial firm and cannot be compared to a national laboratory. Nevertheless, its founding sprang from the realization that science and industrial progress are inseparable ([232], [31]).

2.3.1 Formulae for Radiation Intensity

In 1896 Wilhelm (Willy) Wien[3] published an equation for the intensity of the radiation spectrum emitted by a blackbody. He referenced the earlier work of Eugen von

[3] Wilhelm Wien used the name Willy Wien on the author line of his papers.

Lommel (1837–1899) and Wladimir Michelson (1860–1927), who had failed to obtain a thermodynamic basis for the temperature ([283]; [172]; [191]).

Wien stressed the importance of deriving an expression for the entropy of the molecules interchanging energy with the radiation, then maximizing this entropy to obtain the conditions present in the hohlraum. But there was simply no way to formulate the exchange of energy between the radiation and the molecules.

Michelson had come closest to this. He had applied a Maxwellian distribution to the energy of the molecules comprising the solid walls of the hohlraum and assumed that the period of the radiation produced was inversely proportional to the velocity of the molecule. With this assumption he was able to obtain a formula for the wavelength of the peak in the blackbody spectrum as a function of the thermodynamic (absolute) temperature of the hohlraum as $\lambda_{\mathrm{m}} = \text{constant}/\sqrt{T}$. However, this was not the required relation between the radiation intensity and (λ, T) [191].

Michelson's assumption of the relationship between the radiation intensity and the kinetic motion of the molecules inspired Wien to a similar approach. Instead of the molecules in the wall of the hohlraum, however, Wien considered the gas molecules in the space actually contained in the hohlraum, which had to be present and in thermal equilibrium with the radiation.

The difficulty was that there was no law connecting the motion of the gas molecules and the radiation they would emit. Wien admitted that any assumption relating the motion of a gas molecule to the radiation it emitted was completely arbitrary. So he suggested that the most sensible approach was to make this assumption as simple as possible. He assumed that the wavelength of the radiation emitted by a molecule was a function of the energy of the molecule, and he chose a relationship of direct proportionality [284].

Since the Maxwellian distribution of the molecular kinetic energies (proportional to the square of the velocity) is (see (1.12) in Sect. 1.3.4)

$$v^2 \exp\left(-v^2/\alpha^2\right) ,$$

in which α^2 is a function of temperature, he chose the radiation intensity[4] to be

$$u_\lambda (T) = F(\lambda) \exp(-c_2/\lambda T) . \tag{2.2}$$

Wien produced the argument of the exponential $c_2/\lambda T$ in a separate discussion. He then introduced the Stefan–Boltzmann T^4 law ([258], [21]) for the radiation density by requiring

$$\int_0^\infty F(\lambda) \exp(-c_2/\lambda T)\, \mathrm{d}\lambda = \text{const} \cdot T^4$$

and used integration by variable coefficients to arrive at

[4] Wien used φ_λ to represent the radiation intensity.

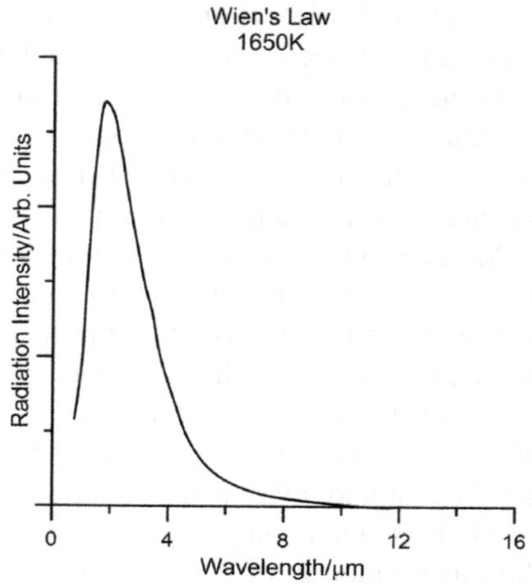

Fig. 2.3 Blackbody radiation spectrum from Wien's law at $T = 1650$ K. Data from Fig. 3 of [175] (Plotted by CSH)

$$u_\lambda(T) = \frac{c_1}{\lambda^5}\exp\left(-\frac{c_2}{\lambda T}\right), \tag{2.3}$$

for the radiation density in the interval $\lambda \rightarrow \lambda + \mathrm{d}\lambda$. [284] Wien's formula fit the low wavelength data very well ([284]; [156]; [157], p. 4). Figure 2.3 shows a plot of the blackbody radiation spectrum from Wien's formula at 1650 K.

Four years later Lord Rayleigh (John William Strutt) (1842–1919) proposed a modification of Wien's formula. The formula Rayleigh proposed is

$$u_\lambda(T) = c_1\frac{T}{\lambda^4}\exp\left(-\frac{c_2}{\lambda T}\right). \tag{2.4}$$

Rayleigh's paper is only two pages long. He was familiar with what was going on in Germany and held Planck's thermodynamics in high regard, but considered Wien's formula to be "little more than a conjecture." He noted, however, that Wien's formula had, in 1900, met with important confirmation. He proposed his modification of Wien's formula by noting that the number of wave vectors in a spherical shell $\mathbf{k} \rightarrow \mathbf{k}+\mathbf{dk}$ is proportional to $k^2 dk$ or, in terms of wavelength, $\lambda^{-4}\mathrm{d}\lambda$, which differed from the $\lambda^{-5}\mathrm{d}\lambda$ that Wien's formula produced. The factor λ^{-4} appears in (2.4) [234]. Figure 2.4 shows a plot of the blackbody radiation spectrum from Rayleigh's formula at 1650 K.

2.3.2 Sources of Radiation

In approaching the second law through the interaction of matter and radiation, Planck believed he could circumvent the problems Boltzmann had encountered with molecules. Planck introduced what he called *natural radiation*, which he formulated

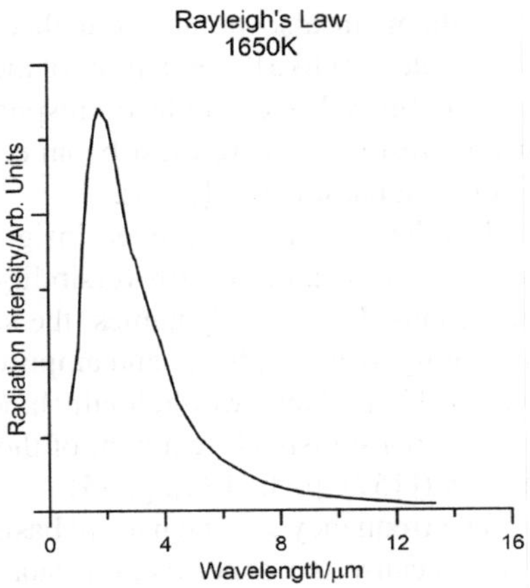

Fig. 2.4 Blackbody radiation spectrum from Lord Rayleigh's law at $T = 1650$ K. Data from Fig. 3 of [175] (Plotted by CSH)

by considering the process of absorption and re-emission of radiation by a blackbody. If thermal radiation from sources at two temperatures is absorbed by a blackbody, the radiation emitted by the body will have the temperature of the body. This implied an irreversible transfer of radiation inside the blackbody. Klein notes that the assumption of natural radiation is equivalent to Boltzmann's hypothesis of molecular chaos ([157], p. 4; [184], p. 36).

For his resonator, which was the source of the interaction between matter and the energy in the radiation field, Planck chose a charged linear oscillator driven by the sinusoidal electric field of the thermal radiation wave with a small "damping term" representing loss of energy to the wave. He then obtained a relationship between the field energy density $u_\nu(T)$ and the average energy of the resonators $U_\nu(T)$ as functions of the radiation frequency ν at the temperature T, which was ([222]; [215], p. 369; [248], pp. 145–147)[5]

$$u_\nu(T) = \frac{8\pi\nu^2}{c^3} U_\nu(T). \tag{2.5}$$

This relationship holds for radiation and resonator motion in the frequency range $\nu \rightarrow \nu + d\nu$.

Because of the presence of a single ν on both sides of (2.5), the mathematical statement is that a resonator with a particular frequency interacts only with radiation at the same frequency. Mathematically this comes about because the integral resulting in $U_\nu(T)$ has an integrand with a sharp maximum at the natural frequency ν. This threatened to destroy the interaction among resonators, which was central to Planck's concept of natural radiation.

[5] Equation (2.5) Is central to the discussion. In [222] Planck obtained the various parts of (2.5) but put them together in terms of frequencies in [225]. Sommerfeld and Pais both derive (2.5).

Boltzmann also pointed out that Maxwell's equations of electrodynamics are reversible, just like the equations of mechanics, and could not result in a description of irreversibility. For example, corresponding to any proposed spherical wave emerging from a resonator there must be an equivalent contracting wave transferring energy to the resonator ([27]; [228]).

But here Planck's deeper insight prevailed. The reversibility of Newtonian mechanics became the irreversibility of Boltzmann's H-theorem through molecular chaos. In electrodynamics, the source of this chaos found its equivalent in the averaging over the phases and amplitudes of the harmonic waves emitted by the resonators [27]. Planck would finally have to face the same issue that caused Boltzmann to turn to a statistical treatment of the entropy of a collection of moving atoms. But not yet ([157], p. 3; [184], p. 35)

The frequency ν is the natural basis for our present (21st century) thinking. However, because the experimental data were gathered in terms of wavelength λ, the original formulae of Wien (2.3) and Rayleigh (2.4), as well as that of Planck, were also presented in terms of wavelength λ. We will therefore continue the discussion in this section in terms of λ. The wavelength and frequency of the thermal radiation are related by $\nu\lambda = c$, where c is the speed of light. Therefore the relationship between the frequency and wavelength intervals $\mathrm{d}\nu$ and $\mathrm{d}\lambda$ is $\mathrm{d}\nu = \left(-c/\lambda^2\right)\mathrm{d}\lambda$, with the minus sign indicating only that ν decreases as λ increases.

Converting Planck's energy relationship (2.5) into a relationship based on wavelength λ, we have

$$u_\nu\left(T\right) = \frac{8\pi\nu^2}{c^3}U_\nu\left(T\right) \to u_\lambda\left(T\right) = \frac{8\pi c^2}{\lambda^2 c^3}U_\lambda\left(T\right)\frac{c}{\lambda^2} \tag{2.6}$$

or

$$u_\lambda\left(T\right) = \frac{8\pi}{\lambda^4}U_\lambda\left(T\right) \tag{2.7}$$

2.3.3 *Thermodynamics and Radiation*

In 1897 Friedrich Paschen (1865–1947) published the results of his measurements in the near infrared ($\lambda = 1$–$8\ \mu$m, $T = 400$–1600 K) showing agreement with Wien's formula. His conclusions regarding Wien's formula were very positive ([218], cited by Pais [215], p. 366).

Planck thus had Wien's formula for the energy density in the radiation field and the relation between that and the energy density of the resonators. Thermodynamics would provide him with the relationships among the internal energy, entropy, and thermodynamic temperature. He was now prepared to undertake an analysis of the interaction between the resonators in the wall of the hohlraum and the radiation field in terms of the second law.

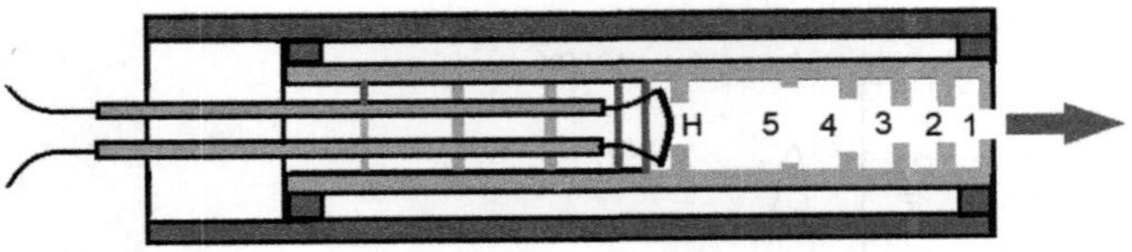

Fig. 2.5 Lummer and Kurlbaum's hohlraum. The hohlraum is in light grey. The hole allowing the passage of radiation is labeled 1 and the diaphrams are labeled 2–5. The electrical heater is *H* ([282] drawn by CSH)

From the Gibbs equation[6] the general definition of the thermodynamic temperature T is

$$\left(\frac{\partial S_\lambda}{\partial U_\lambda}\right)_V = \frac{1}{T}, \tag{2.8}$$

where S_λ is the entropy of the resonators[7] in the wavelength interval $\lambda \to \lambda + d\lambda$. Because the volume of the metal containing the resonators is constant, the partial derivative in (2.8) is simply the full derivative dS_λ/dU_λ.

The actual apparatus used to make the measurements on which Planck based his theory was designed by Otto Lummer (1860–1925) and Ferdinand Kurlbaum (1857–1927) in 1898. The Lummer and Kurlbaum hohlraum was a cylindrical platinum box with a hole in the end. Internally, the box was divided by diaphragms and was blackened with iron oxide [174]. Figure 2.5 is a simplified picture of Lummer and Kurlbaum's hohlraum. The hole for emitting radiation is at position 1. The diaphragms are at positions 2–5. *H* is the electrical heater.

In Fig. 2.6 we have plotted the general form of the blackbody radiation spectrum, which is the radiation intensity versus wavelength. The *standard units* of intensity are $\mathrm{kW\,Sr^{-1}m^{-2}nm^{-1}}$ and the units of wavelength are μm.

The maximum temperature considered in the period of interest to us was 1650 K. At $T = 5000$ K, the peak of the blackbody spectrum is in the visible region. Therefore, essentially the entire blackbody spectrum at 1650 K is in the infrared region. In the last decade of the nineteenth century it was still particularly difficult to measure radiation in this region, and the problem was only just being resolved.

By 1850, it had become possible to measure radiation with a wavelength of $\lambda \approx 1.5\ \mu$m, but the near infrared region begins at 0.78 μm. Then, during a period spent in Berlin, the American physicist Ernest Fox Nichols (1869–1924) discovered an anomalous reflection of radiation from polished quartz. Near $\lambda = 9\ \mu$m, the reflectivity changed rapidly from a few percent to almost the reflectivity of polished silver. The absorption and hence the reflection of radiation from crystals is dependent

[6] The Gibbs equation

$$TdS = dU + PdV$$

results from a mathematical combination of the first and second laws of thermodynamics ([138], p. 29).

[7] We note that this is the entropy of the oscillators, not of the radiation field.

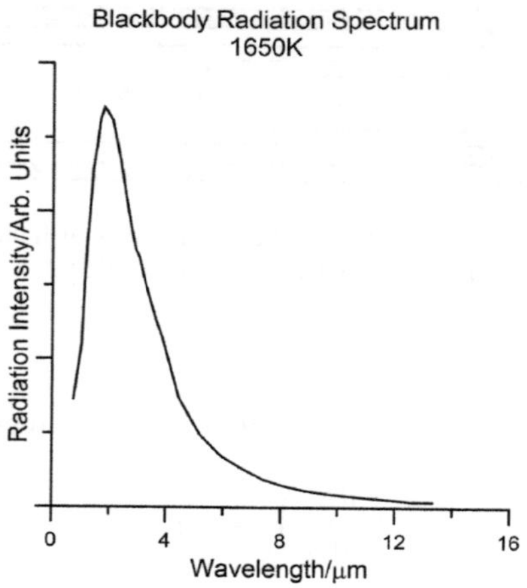

Fig. 2.6 Blackbody radiation spectrum recorded at $T = 1650$ K. Data from Fig. 3 of [175] (Plotted by CSH)

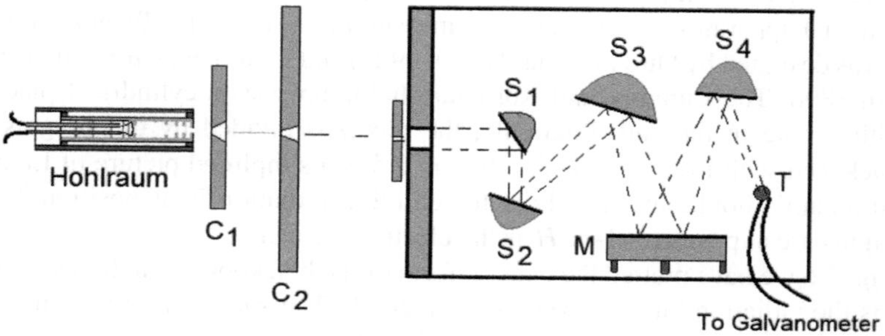

Fig. 2.7 Hohlraum with collimator and detector. C_1 and C_2 form the collimator providing a narrow beam from the hohlraum entering the detector chamber. The surfaces S_1–S_4 are the crystals from which the residual rays are reflected. M is a concave mirror to focus the beam onto the final crystal S_4. T is a thermopile providing the electrical signal to the very sensitive galvanometer ([239] drawn by CSH)

on the wavelength of the radiation, and for some crystals quite dramatically so. This is now understood in terms of lattice vibrations.

In the years following the initial quartz studies, Nichols and Heinrich Rubens (1865–1922) found long wavelength, narrow band reflections of this sort in a number of ionic crystals. Combinations of these crystals in series could then be used to obtain very nearly monochromatic beams in the infrared ([153]; [207]; [208]; [215], p. 366).

Figure 2.7 presents a schematic diagram of the sort of apparatus that was used by Rubens and Kurlbaum. The hohlraum used to produce the thermal radiation was of the Lummer and Kurlbaum type shown in Fig. 2.5. The diaphragms C_1 and C_2 constituted the collimator for the beam. The collimated beam entered a box in which were mounted fluorite (CaF_2) or rocksalt ($NaCl$) crystals S_1–S_4 and a concave mirror

Fig. 2.8 Main building (Hauptgebäude) of the RWTH Aachen (Creative Commons Attribution-Share Alike 2.5 Generic license. Image by Aleph)

silvered on the front to focus the beam onto S_4 and then the thermopile at T. The current from the thermopile went to a very sensitive galvanometer [239].

Wien was a colleague of Planck's in Berlin until he took a position at the Technical University (*Rheinisch-Westfälische Technische Hochschule*: *RWTH*) in Aachen[8] in 1896. Figure 2.8 is a modern photograph of the main building (Hauptgebäude) of the RWTH Aachen. Planck was thus well acquainted with Wien's work, but he wanted to avoid the labor of considering molecular motion, basing his treatment on thermodynamics alone ([184], p. 31). On the other hand, he was interested in obtaining the thermodynamics behind Wien's formula, since this formula fit the data in the short wavelength region. Here we follow what was probably Planck's line of reasoning.

For convenience, we repeat Wien's formula for the radiation spectrum (2.3):

$$u_{\lambda}(T) = \frac{c_1}{\lambda^5} \exp\left(-\frac{c_2}{\lambda T}\right). \tag{2.9}$$

Using (2.7), the energy density of the resonators producing the short wavelength range of the radiation is

$$U_{\lambda}(T) = \frac{c_1}{8\pi\lambda} \exp\left(-\frac{c_2}{\lambda T}\right). \tag{2.10}$$

Solving (2.10) for $1/T$ and using (2.8), we have

---

[8] Rheinisch-Westfälische Technische Hochschule Aachen is a research institute located in Aachen, North Rhine-Westphalia, Germany.

$$\frac{dS_\lambda}{dU_\lambda} = \frac{\lambda}{c_2} \ln\left(\frac{c_1}{8\pi\lambda}\right) - \frac{\lambda}{c_2} \ln U_\lambda. \tag{2.11}$$

Integrating (2.11) yields the entropy of the resonators in equilibrium with the short wavelength portion of the radiation field, viz.,

$$S_\lambda = -\frac{\lambda U_\lambda}{c_2} \ln\left(\frac{8\pi\lambda U_\lambda}{c_1 e}\right), \tag{2.12}$$

where e is the base of the natural logarithm.

Identifying

$$\beta = \frac{c_2}{c}, \quad a = \frac{c_1}{8\pi c},$$

Equation (2.12) takes the form

$$S_\lambda = -\frac{\lambda U_\lambda}{\beta c} \ln\left(\frac{\lambda U_\lambda}{cae}\right), \tag{2.13}$$

which can be written as

$$S_\nu = -\frac{U_\nu}{\beta\nu} \ln\left(\frac{U_\nu}{ae\nu}\right). \tag{2.14}$$

Equations (2.13) and (2.14) were both cited by Planck for the entropy of the resonators in his extensive paper on irreversible radiation processes. This paper dealt with entropy transport by radiation and resonators, and Planck showed that Wien's formula was consistent with the requirements of entropy maximum at equilibrium [222].

The condition for thermodynamic stability at equilibrium is that the entropy should be a maximum under conditions of constant (U, V) (see e.g. [138], Chap. 12). Therefore, at constant volume, the curvature of the projection of S onto a plane of constant V should be negative, i.e., $\left(\partial^2 S/\partial U^2\right)_V < 0$. In general, for systems in which the volume is constant,

$$\frac{d^2 S_\lambda}{dU_\lambda^2} = \text{any negative function of } U_\lambda. \tag{2.15}$$

Using (2.11), we see that, for Wien's formula,

$$\frac{d^2 S_\lambda}{dU_\lambda^2} = -\frac{\lambda}{c_2}\frac{1}{U_\lambda} < 0, \tag{2.16}$$

as required for thermodynamic stability. Expressing (2.16) in a more general form, we obtain

$$\frac{d^2 S_\lambda}{dU_\lambda^2} = -\frac{g(\lambda)}{U_\lambda}, \tag{2.17}$$

where (2.16) implies $g(\lambda) = \lambda / c_2$ [224].

2.3.4 Measurements in the Infrared

On 18 May 1899 Planck presented this derivation of Wien's law to the Berlin Academy [222], and on 7 November he sent the same paper to *Annalen der Physik*. In the November paper, Planck called for further measurements. When he received the proofs of the paper the experiments he had called for were already underway, and he was therefore able to note in proof that these results were not consistent with Wien's law.

The problem is clearly visible in the data presented by Lummer and Ernst Pringsheim (1859–1917) at the meeting of the *Deutsche Physikalische Gesellschaft* (DPG, German Physical Society) on 2 February 1900. Figure 2.9 shows a plot of their data as it appeared in the proceedings of the DPG [175].The data obtained by Lummer and Pringsheim lie on the solid black line. The Rayleigh and Wien's laws result in the dashed red and green lines. Both laws show good agreement for short wavelengths, but substantial deviation in the long wavelength regime.

To discuss fits to their data Lummer and Pringsheim introduced a generalized formula that incorporated Rayleigh's and Wien's laws and the result of some of

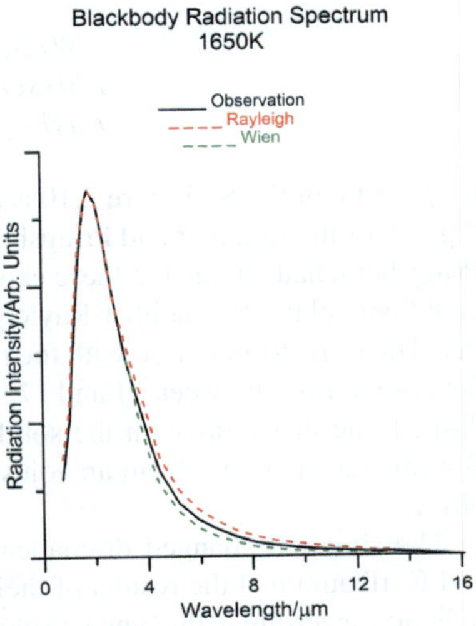

Fig. 2.9 Blackbody radiation spectra at 1650 K. The data are on the solid black line flanked on both sides by the predictions from Rayleigh's and Wien's laws. Data from Fig. 3 of [175] (Plotted by CSH)

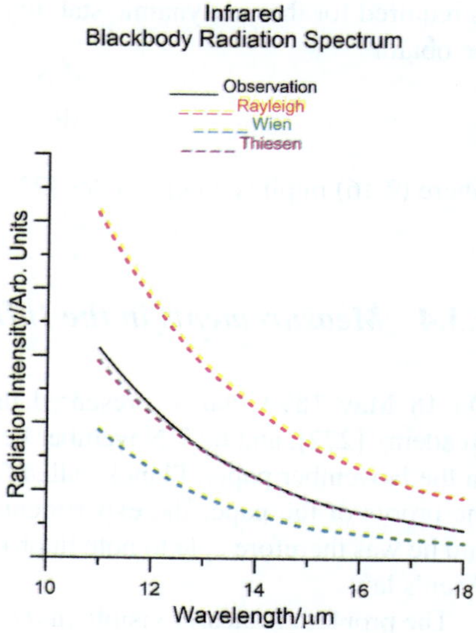

Fig. 2.10 Lummer and Pringsheim (18 μm) indicating deviations from laws of Rayleigh and Wien and possible agreement with a fit based on ideas of Thiesen. Data from Fig. 3 of [175] (Plotted by CSH)

the general ideas of Max Thiesen (1849–1936), given in a presentation at the same meeting [265]. This generalized formula was

$$u_\lambda(T) = C_1 T^{5\text{-}\mu} \lambda^{\text{-}\mu} \exp\left(-\frac{C_2}{(\lambda T)^\nu}\right). \tag{2.18}$$

The parameters μ and ν were introduced by Lummer and Pringsheim. From (2.18), one can obtain the laws of Wien, Thiesen, and Rayleigh by the substitutions

$$\begin{aligned} Wien &: \mu = 5,\ \nu = 1\,, \\ Thiesen &: \mu = 4.5, \nu = 1\,, \\ Rayleigh &: \mu = 4,\ \nu = 1\,, \end{aligned}$$

for μ and ν in (2.18). Figure 2.10 is a copy of the graph that appeared as an insert in Fig. 2.9 of the Lummer and Pringsheim paper. The longest wavelength Lummer and Pringsheim had attained in these experiments was 17.9 μm. At this wavelength they could only claim that neither Rayleigh nor Wien's law agreed with observation, but that Thiesen's formulation with the parameters $\mu = 4$ and $\nu = 1.3$ matched most of the observations between 11 and 17.9 μm. The Thiesen formula produced the purple dashed line that joins with the solid black line beyond about 12 μm in Fig. 2.10. But this resulted only from an adjusting of parameters and revealed nothing of the physics.

The situation changed dramatically in October of 1900. By that time Rubens and Kurlbaum had the results of their very careful measurements of the blackbody radiation spectrum in the long wavelength regime. These firmly established that there was considerable disagreement with Wien's formula, as Planck had anticipated in his note in proof. The data Rubens and Kurlbaum were gathering included measurements at $\lambda = 24.0$, 31.6, and 51.2 μm, which represented a tour de force in experimental physics. They were preparing a paper for the meeting of the DPG that would take place on 19 October 1900.

According to Planck's student Gerhard Hettner (1892–1968), on Sunday 7 October 1900 Rubens and his wife visited the Plancks. In the course of the afternoon's discussion, Rubens said that the results of his measurements with Kurlbaum indicated that the formula proposed by Rayleigh (2.4) was valid in the long wavelength regime, while Wien's formula (2.3) was not [234]. In any case it was clear that the correct formula for blackbody radiation had to go over into the form Rayleigh proposed for high values of λT [141].

Planck had already obtained (2.16) for Wien's formula, which was valid for short wavelengths but, as he now realized, failed for long wavelengths. And Rubens had just told him that for large values of λ the formula Rayleigh had just published for the radiation density in the long wavelength regime matched the data he and Kurlbaum were gathering. We can only guess the thoughts that went through Planck's mind while he continued as host during that Sunday afternoon.

Hettner does not tell us that, as soon as the Rubenses left, Planck retired immediately to his study to begin work on finding a new formula. However, Hettner did say that on that same evening Planck sent a postcard to Rubens with a new formula on it. And one or two days later, Hettner writes, Rubens went back to Planck to tell him that his formula agreed point by point with the long wavelength measurements. Then the following Friday, 19 October, Kurlbaum reported the long wavelength measurements he and Rubens had made to the meeting of the DPG. After that Planck presented his most recent formula and the fact that the measurements Kurlbaum had just reported matched this formula. He also included some numerical examples ([141]; [224]).

We cannot be certain of the steps Planck followed to obtain his formula. We know only that his reasoning was thermodynamically based, that he was very familiar with the problem, and that he had just been given the news that Rayleigh's formula matched the data for long wavelengths. From these snippets of information we can attempt to reconstruct a possible path, which we may suppose followed the approach he had used to obtain Wien's formula for low values of λ.

Rayleigh's formula is (2.4), which we repeat here for convenience:

$$u_\lambda(T) = c_1 \frac{T}{\lambda^4} \exp\left(-\frac{c_2}{\lambda T}\right). \tag{2.19}$$

Using (2.7), the energy of the resonators is

$$U_\lambda(T) = c_1 \frac{T}{8\pi} \exp\left(-\frac{c_2}{\lambda T}\right). \tag{2.20}$$

Planck was only interested in this relationship for large values of λT, since this was the only region in which Rayleigh's formula fit the data. For large values of λT, the exponential in (2.20) approaches unity. Then

$$U_\lambda(T) = \frac{c_1}{8\pi} T. \tag{2.21}$$

To obtain an equation for $\mathrm{d}S_\lambda/\mathrm{d}U_\lambda$, we turn once again to the expression for $1/T$ obtained from (2.21). This is

$$\frac{1}{T} = \frac{\mathrm{d}S_\lambda}{\mathrm{d}U_\lambda} = \frac{c_1}{8\pi}\frac{1}{U_\lambda}. \tag{2.22}$$

We may integrate (2.22) to obtain the entropy for large λT as

$$S_\lambda = \frac{c_1}{8\pi} \ln U_\lambda. \tag{2.23}$$

The more critical term is the second derivative $\mathrm{d}^2 S_\lambda/\mathrm{d}U_\lambda^2$, since this determines whether or not the proposed system of resonators is thermodynamically stable. By differentiating (2.22), we find

$$\frac{\mathrm{d}^2 S_\lambda}{\mathrm{d}U_\lambda^2} = -\frac{c_1}{8\pi}\frac{1}{U_\lambda^2} \tag{2.24}$$

for large values of λT. This result satisfies the thermodynamic stability requirement in (2.15). Therefore Raleigh's formula represents a system of resonators that is thermodynamically stable.

Although four years later Einstein (1879–1955) would show that U^2 is proportional to the statistical average of the square of the fluctuation of the resonator energy[9] $\langle \delta U^2 \rangle$, in 1900 Planck had no physical understanding of the origin of the U^2 term in $\mathrm{d}^2 S_\lambda/\mathrm{d}U_\lambda^2$ ([94] referenced by [215], p. 69).

In his obituary for Planck, Max Born (1882–1970) wrote (2.24) in the form

$$\frac{\mathrm{d}^2 S_\lambda}{\mathrm{d}U_\lambda^2} = -\frac{C}{U_\lambda^2}, \tag{2.25}$$

and (2.16) in the form

$$\frac{\mathrm{d}^2 S_\lambda}{\mathrm{d}U_\lambda^2} = -\frac{1}{BU_\lambda}. \tag{2.26}$$

---

[9] The system energy F is equal to its statistical average $\langle F \rangle$ plus a fluctuation δF, i.e.,

$$F = \langle F \rangle + \delta F,$$

where $\langle \delta F \rangle = 0$. However,

$$\langle \delta F^2 \rangle = \langle F^2 \rangle - \langle F \rangle^2.$$

The problem for Planck, as Born put it, was to combine these two limiting cases into one. According to Born, Planck noticed he could do this if he first considered the reciprocal of $\mathrm{d}^2 S_\lambda/\mathrm{d}U_\lambda^2$ and added the two equations [27]. That is,

$$\left(\frac{\mathrm{d}^2 S_\lambda}{\mathrm{d}U_\lambda^2}\right)^{-1} = -\left(BU_\lambda + \frac{1}{C}U_\lambda^2\right). \tag{2.27}$$

Then,

$$\frac{\mathrm{d}^2 S_\lambda}{\mathrm{d}U_\lambda^2} = -\frac{C}{U_\lambda\,(U_\lambda + BC)}. \tag{2.28}$$

Hettner refers to this as a bold and extremely fortunate thought (glückliche Gedanken). [141]

Integrating (2.28) and using (2.8), Planck had

$$\frac{\mathrm{d}S_\lambda}{\mathrm{d}U_\lambda} = \frac{1}{T} = \frac{1}{B}\ln\frac{BC + U_\lambda}{U_\lambda}, \tag{2.29}$$

which is

$$U_\lambda = C\frac{B}{\left[\exp\left(B/T\right) - 1\right]}. \tag{2.30}$$

The terms B and C in (2.30) are independent of U_λ, but otherwise arbitrary.

Using (2.7) in (2.30) yields the radiation intensity

$$u_\lambda = \frac{8\pi}{\lambda^4}C\frac{B}{\exp\left(B/T\right) - 1}. \tag{2.31}$$

To guarantee agreement with Rayleigh's formula for large values of λT, we replace replace B with b/λ in (2.31):

$$u_\lambda = \frac{8\pi}{\lambda^5}C\frac{b}{\exp\left(b/\lambda T\right) - 1}. \tag{2.32}$$

If we now combine $8\pi bC$ into a single constant C, we have the formula Planck presented at the DPG meeting on 19 October 1900, which is

$$u_\lambda = \frac{C}{\lambda^5}\frac{1}{\exp\left(b/\lambda T\right) - 1}. \tag{2.33}$$

Planck denoted our b by c, but we shall not do this because c is usually reserved for the speed of light in vacuum.

2.4 Probability and the Quantum

Although the formula Planck presented to the DPG on 19 October 1900 matched the long and short wavelength data exactly, it was still only an empirical fit to those data based on a guess. The guess had the weight of experience and deep physical insight, but it was nevertheless still a guess. Planck had been working on the problem of matter and radiation in equilibrium with limited success since 1894. But now that he had the correct formula for the entire spectrum, the next issue was to understand the physics that resulted in that formula. After his rather dramatic presentation of his formula at the October DPG meeting, Planck reached what he later referred to as a state of despair. There were just under two months before the December meeting of the DPG and he realized that he would have to explain the physical understanding of his formula.

In a letter from 1931 he wrote, "a theoretical interpretation had to be found at any cost no matter how high." And then, "After a few weeks of the most strenuous work of my life, the darkness lifted and an unexpected vista began to appear." (quoted from a letter Planck wrote to Robert W. Wood (1868–1955) [230] cited in [156]) The lifting of the darkness came with Planck's acceptance of Boltzmann's statistical ideas in the formulation of the entropy of the resonators and his subsequent discovery of the energy quantum.

As Klein pointed out, Planck noticed that Boltzmann's H was based on a logarithm, as were the relationships Planck had found for the entropy in both the low λ limit (2.13) and the high λ limit (2.23). And so, perhaps with some reluctance, but with much determination, Planck went back to Boltzmann's 1877 paper ([20]; [156])

In this paper, Boltzmann was still seeking a deeper understanding of the H-theorem. In fact, he dealt with it in three of the five sections of the paper, although he did not use the symbol H to denote the term he was studying. His primary interest was in molecular conditions in a gas at equilibrium. In the final section[10] he called this term Ω and showed that at equilibrium it would yield the *Maxwell* distribution function

$$f\left(\mathbf{r},\mathbf{v}\right)=\frac{N}{V\left(4\pi T/3m\right)^{-3/2}}\exp\left(-\left(3m/4T\right)v^{2}\right). \tag{2.34}$$

Here we have maintained Boltzmann's original notation in which T had the units of energy. What would later become the Boltzmann constant does not appear here. With the distribution function (2.34), Boltzmann showed by direct calculation that the entropy, which he wrote as $\int \mathrm{d}Q/T$, was equal to a constant times Ω. Because of the units he was using, this constant was $2/3$. It would later become the Boltzmann constant.

In the first section of the paper Boltzmann treated the problem of finding the equilibrium conditions for the case in which the energy of the system was divided into parcels. This was a mathematical simplification. He also considered such a division to

[10] This is section V of the paper. The title of the section is *Relationship of the Entropy to that Quantity Which I Have Called the Probability Distribution*.

be unphysical. In the second section he solved the problem for continuous energies, obtaining the now familiar variational problem with two undertermined Lagrange multipliers. In each section, he carried out the mathematics in great detail. Although the paper was long, Planck only really needed the first section. Indeed, Planck used Boltzmann's notation in that first section for his presentation to the DPG on 14 December.

Planck first presented his understanding of the physics at the DPG in the meeting of 14 December 1900 [225]. This presentation was then followed by a publication in *Annalen der Physik* in 1901 [226]. There was no change in the physics in these two publications, although the 1901 publication gave a more detailed description for a broader audience.

Planck had come to a new understanding of the meaning of entropy based on Boltzmann's concept of entropy as a statistically based quantity. He recognized and used Boltzmann's term "complexion" to designate an arrangement of the energy elements among the resonators in the hohlraum walls. When he stood before his colleagues at the DPG meeting Planck knew that the number of energy elements was finite and he could easily calculate the number of possible complexions among N resonators and P energy elements. This number he designated as $\mathfrak{K}$, which is K in the German Fraktur script (𝔨𝔬𝔪𝔭𝔩𝔢𝔵𝔦𝔬𝔫 in German Fraktur). He denoted the maximum number of complexions for a given total resonator energy $\mathcal{E}_0$ by $\mathfrak{K}_0$. He then identified the entropy of the resonators as $k \log \mathfrak{K}_0$, where $k = 1.346 \times 10^{-16}$ erg K^{-1}, which was respectably close to the modern value of $1.3806568 \times 10^{-16}$ erg K^{-1} for Boltzmann's constant.

In the 1901 paper Planck denoted the entropy by[11]

$$S = k \log W + \text{constant}. \tag{2.35}$$

This shows that Planck understood rather clearly what Boltzmann never quite stated in the 1877 paper: the entropy is equal to a constant times $\log W$. This was demonstrated more directly in the first section of Boltzmann's paper than the last. Planck's form of the equation, without the constant that Planck dropped, is the equation appearing on Boltzmann's tombstone in the *Wiener Zentralfriedhof* (Vienna Central Cemetery).[12]

The positive sign in (2.35) rather than the negative sign relating entropy to the H-function (1.16) resulted from the fact that W, although it stood for probability (*Wahrscheinlichkeit*), was the number of complexions, hence a large number, while the distribution function in H was less than unity. The actual relationship between the entropy and H, and certainly between the entropy and the logarithm of the number of complexions, is not trivial, as was subsequently shown by Edwin T. Jaynes (1922–

[11] We use log rather than ln as the designation for the natural logarithm, because that is the notation used by Planck in the original.

[12] According to Arnold Sommerfeld (1868–1951), this equation was first written by Planck in 1906, in the first edition of his *Vorlesungen über die Theorie der Wärmestrahlung* ([248], pp. 213–220). However, it is also Eq. (3) in Planck's 1901 paper.

1998) [149]. These issues illustrate the radical nature of certain steps required to make progress along the road to theoretical physics.

Planck began the 1901 *Annalen der Physik* paper by stating that the experimental results of Lummer and Pringsheim [175] and those of Rubens and Kurlbaum [239] had shown that both Wien's and his own work, based on a theory of radiation, had no general validity. He then outlined the approach that he would be using:

- The entropy would be required as a function of the resonator energy, i.e., $S = S(U)$.
- The functional dependence of the resonator energy on temperature, $U = U(T)$, could be found from $S = S(U)$ and the definition of the thermodynamic temperature, viz., $\mathrm{d}S/\mathrm{d}U = 1/T$.
- The radiation energy density, $u = u(T)$, could be found from the relationship between the radiation density u and the resonator energy U, viz., $u = \left(8\pi\nu^2/c^3\right) U$.

The entire problem then depended on finding the entropy of the resonators as a function of the resonator energy. For this, he wrote, a new understanding of the meaning of entropy would be required.

Planck began by discussing time scales and, implicitly, the meaning of measurement. The time interval to be considered, he wrote, would be long compared to the period of oscillation of a resonator, but small compared to the time of any physical measurement. Then he pointed out that there was an irregularity (*Unregelmässigkeit*) in the manner in which a resonator constantly changes its amplitude and phase. If the amplitude and phase were constant it would be possible for vibrational energy to be completely converted into work.[13] If this could be accomplished, there would be no entropy. The constant energy of a single stationary resonator could then only be defined as a time average, or equivalently as the average over a large number N of identical resonators. These identical resonators would be subjected to a stationary radiation field and far enough apart to have no effect on one another.

He could then speak of the total energy of the N resonators as U_{N}, with

$$U_{\mathrm{N}} = NU, \tag{2.36}$$

where U was the energy of a single resonator. This system of resonators then had a total entropy S_{N}, where

$$S_{\mathrm{N}} = NS, \tag{2.37}$$

allowing him to speak of the entropy of a single resonator as S. This total entropy S was a result of the lack of order in the manner in which the energy was distributed among the resonators.[14]

---

[13] Planck provided no additional support for this contention, although his knowledge of electrodynamics and matter was unsurpassed. Complete conversion of energy into work would violate the second law of thermodynamics.

[14] A lack of order in the distribution of energy quanta among resonators does not imply a lack of order among the resonators of the system.

Planck then turned to $S = k \log W$ with the additive constant dropped and asked for the meaning of W, because there was no mention of probability in the laws of electrodynamics. He noted that, for expediency and simplicity, one could exploit the similarity with the problem of the kinetic gas theory. The energy of the N resonators U_{N} would then be divided into a very large number P of energy elements ε, so that

$$U_{\mathrm{N}} = P\varepsilon. \tag{2.38}$$

For the time being Planck left ε undefined. He then pointed out that it was completely clear that the number of ways of distributing the P energy units of magnitude ε among the N resonators was finite. This number was the number of complexions $\mathfrak{R}$. From combinatorics, he knew that this number was

$$\mathfrak{R} = \frac{(N+P-1)!}{(N-1)!P!} = \frac{(N+P)^{(N+P)}}{N^N P^P}, \tag{2.39}$$

where the second equality came from Stirling's formula, which Planck took to be exact because of the very large numbers involved. He then assumed that each complexion was equally probable, pointing out that the validity of this assumption could only be verified by experiment.

From (2.35), dropping the additive constant as Planck did, the entropy of the system was

$$\begin{aligned} S_{\mathrm{N}} &= k \log \mathfrak{R} \\ &= k\left[(N+P)\log(N+P) - N\log N - P\log P\right] \\ &= Nk\left[\left(1+\frac{U}{\varepsilon}\right)\log\left(1+\frac{U}{\varepsilon}\right) - \frac{U}{\varepsilon}\log\left(\frac{U}{\varepsilon}\right)\right], \end{aligned} \tag{2.40}$$

since $P/N = U/\varepsilon$. Dropping the factor N in (2.40), the entropy of a single resonator, as defined by Eq. (2.37), was thus[15]

$$S = k\left[\left(1+\frac{U}{\varepsilon}\right)\log\left(1+\frac{U}{\varepsilon}\right) - \frac{U}{\varepsilon}\log\left(\frac{U}{\varepsilon}\right)\right]. \tag{2.41}$$

The issue was now to obtain a formula for the ratio U/ε. Planck knew that his own formula (2.33) matched Wien's for short wavelengths, so he could use any consequences of Wien's formula for short wavelengths. This provided a way to find the ratio U/ε.

At the DPG meeting on 2 February 1900 Thiesen had presented a paper considering the consequences of Wien's formula for the radiation density. He began with this formula in the form

[15] Entropy is a system property, not a property of individual resonators. Therefore, Planck's treatment here of individual resonator energy and entropy must be seen as purely formal, based on his definitions of U and S.

$$u_{\lambda}(T) = T^{5}\psi(\lambda T), \tag{2.42}$$

where $\psi(\lambda T)$ is a function only of the product λT. [265] Beginning with (2.42), Planck was able to show, after some steps, that this resulted in a formulation for the entropy of the resonators as a function of the ratio U/ν alone. That is

$$S = f\left(\frac{U}{\nu}\right). \tag{2.43}$$

Comparing (2.41) with (2.43), Planck concluded that $\varepsilon \propto \nu$, which he wrote as

$$\varepsilon = h\nu, \tag{2.44}$$

where h was a constant to be determined by experiment. Using (2.44) in (2.41), the inverse of the thermodynamic temperature is

$$\frac{1}{T} = \frac{\mathrm{d}S}{\mathrm{d}U} = \frac{k}{h\nu}\log\left(1 + \frac{h\nu}{U}\right). \tag{2.45}$$

Solving (2.45) for the resonator energy U, we have

$$U = \frac{h\nu}{\exp(h\nu/kT) - 1}, \tag{2.46}$$

and using (2.6), Eq. (2.46) becomes

$$u_{\nu} = \frac{8\pi h\nu^{3}}{c^{3}}\frac{1}{\exp(h\nu/kT) - 1}. \tag{2.47}$$

This is the form of the radiation energy density Planck presented in Eq. (12) of his Annalen der Physik (1901) paper.

He then followed this with the wavelength form of the result, viz.,

$$u_{\lambda} = \frac{8\pi ch}{\lambda^{5}}\frac{1}{\exp(ch/k\lambda T) - 1}, \tag{2.48}$$

which is equation (13) of the paper. This is also the form (2.33) that Planck presented to the DPG on 19 October 1900.

In the final section of the paper Planck presented numerical values for a number of constants. The only ones that will interest us here are

$$\begin{aligned} h &= 6.55 \times 10^{-27}\ \text{erg s}, \\ k &= 1.346 \times 10^{-16}\ \text{erg K}^{-1}. \end{aligned} \tag{2.49}$$

Planck's result matched the experimental blackbody radiation results exactly.

There is something important to note here. Theoretical structures are based on time averages of measurable quantities, as Planck made clear. As we pointed out in a footnote above, this excludes fluctuations, which must average to zero. But averages of products of fluctuations are measurable quantities. Even though there may be no practical method to measure these directly, they may produce results which are measurable. The missing correlations we pointed to in the model for collisions in the H-theorem are averages of products of fluctuations in the number densities of the colliding particles, i.e., $\langle \delta N_1 \delta N_2 \rangle$. These missing correlations are the source of the statistical dependence of $H(t)$ that we noted in our plot of $H(t)$ in Fig. 1.6.

As we pointed out above, Einstein would later identify the term U_λ^2 as being proportional to the value of $\langle \delta U^2 \rangle$ for the resonators. Although this is fundamentally no surprise, it indicates the great importance of any fluctuations in the distribution of energies among the resonators. These energy fluctuations are also present in a state of thermodynamic equilibrium. It was this term that was the source of Planck's great discovery, and his state of despair.

Planck pointed to fluctuations when considering the irregular way in which a resonator constantly changes its amplitude and phase. In his mathematical treatment of the system of resonators at equilibrium, this irregularity appeared in the distribution of the energy elements ε among the N resonators. This, according to Planck, was the source of entropy. This identification was central to his conversion to Boltzmann's position on entropy. It seems to the present author that he may even have gone beyond Boltzmann's understanding.

We should note, however, that Planck did not in any way consider the possible electrodynamical or mechanical basis for this microscopic irregularity. Planck's mathematical contention was simply that each complexion was equally probable and that the distribution which had the greatest number of complexions was the one at equilibrium. This implied an exchange of elemental energies among resonators. But Planck did not speculate on that implication. Einstein would later be very explicit about the dynamics.

2.5 Planck's Nobel Lecture

In 1920 Planck was awarded the Nobel Prize in Physics for *The Genesis and Present State of Development of the Quantum Theory*. We include a brief outline of some points in his Nobel Lecture here because they provide us with his own assessment of what happened.

He recognized that his original hope of finding a classical solution had been blocked by the behavior of the resonators he had chosen to study. Kirchhoff had shown that the spectrum of blackbody radiation is a universal function, dependent only upon temperature and wavelength, and in no way upon the properties of any substance producing the radiation This was of great importance, allowing Planck to concentrate on the resonators producing the radiation rather than on the radiation itself. A complex system, composed of many degrees of freedom, was then reduced to a simple system with one degree of freedom.

Planck's original hope had been based on separating the absorption and the emission of radiation by the resonators. However, the resonator only absorbed the rays that it emitted and was totally insensitive to other regions of the spectrum. This had been at the heart of Planck's concept of natural radiation. This was also the point that Boltzmann had raised in his criticism of Planck's concept of irreversibility. Here Planck acknowledged Boltzmann's "riper experience" in these questions.

Then Planck said he turned to thermodynamics and sought a relationship between energy and entropy, using the requirement that the algebraic sign on second derivative d^2S_ν/dU_ν^2 must be negative and that this second derivative had a direct physical meaning. At that time (1899) he was committed to a phenomenological approach and so turned to what was then the accepted representation of the spectrum: Wien's law (2.3). This resulted in (2.15) and (2.17). Then he encountered the long wavelength measurements by Lummer and Pringsheim, and finally by Rubens and Kurlbaum. This produced the term U_ν^2 on the right-hand side of (2.27).

He admitted that the resulting formula served only as a happily chosen interpolation formula. The next step involved seeking a connection between entropy and probability, which was Boltzmann's approach. Then, after the most strenuous work of his life he said, "light came into the darkness, and a new undreamed-of perspective opened up before me."

The remainder of the lecture was devoted to the relationship between the quantum and work by others that had followed his discovery of the quantum, then the work that was ongoing in 1920 [228].

2.6 Summary

In this chapter we have taken a major step forward in our study of the origins of the quantum theory. At this time, Thomson was developing his atomic model with its many rings of electrons immersed in a positive cloud of charge.

Although we have dealt primarily with Planck, we have seen how his ideas were highly dependent on ideas put forward by others, such as Wien, Thiesen, and Rayleigh. He chose to link his theoretical investigations directly to the experimental work at the PTR. His idea, based on Kirchhoff's proof, of concentrating on imaginary resonators was critical. He was then able to deal directly with the relationship between the energy of the resonators and the radiation. This gave him a way to deal with the thermodynamics of the material in the walls of the hohlraum.

We saw that the decisive factor was the long wavelength data from Rubens and Kurlbaum that he was shown that Sunday afternoon in October. With those data and the fact that they agreed with Rayleigh's formula, he was able to use the process he had formulated, along with an inspired guess, to obtain the correct formula for the radiation spectrum.

The next step was the one with which we are at least anecdotally most familiar. Planck turned to Boltzmann's ideas, with which he was already familiar, but had dismissed as incorrect. He came to the realization that the entropy of a system is a function of the ways in which the energy can be distributed among the resonators: the number of complexions. The distribution of the energy quanta among the resonators (fluctuations in the energy distribution) resulted in fluctuations in the entropy. This was a major revision in Planck's understanding of physics. As we have seen in Planck's Nobel Lecture, he came to a deep appreciation of Boltzmann's position and even extended it.

Chapter 3
Electrodynamics and Matter

Everything should be made as simple as possible, but not simpler.
Albert Einstein

3.1 Introduction

By the beginning of the 20th century the classical physics of analytical mechanics and electrodynamics had been extremely successful in providing an understanding of the universe. There was little expectation that this would change radically as a result of a rather small discovery regarding the parcelling of energy in thermal radiation. In this chapter we shall consider the thoughts of two theoretical physicists regarding what we now realize was the beginning of a conceptual revolution. We separate this discussion from our considerations of atomic physics, which we consider in the next chapter.

The physicists we have selected are Planck and Einstein, in part because they represent extremes in their response to the quantum theory. On the one hand we have Planck, who was still not certain how to formulate an understanding of his creation. And on the other hand we have Einstein, who seemed to have no difficulty whatsoever with the quantum, although he did see some quite serious difficulties with Maxwell's electrodynamics and the treatment of atoms, molecules, and electrons. Note also that, although it is of no relevance to the present chapter, Planck and Einstein were personal friends (see, e.g., [39], pp. 210–211).

Because of the importance of Einstein's work, we have devoted considerable space to his development of these ideas in his paper on the photon.

C. S. Helrich, *The Quantum Theory—Origins and Ideas*, History of Physics,
https://doi.org/10.1007/978-3-030-79268-8_3

3.2 Planck's Second Theory

As an indication that Planck had an understanding of the importance of his discovery many authors point to the recollections of Planck's son Erwin (Erwin Planck[1] (1893–1945)). Erwin recalled that, on a walk with his father in the residential area of Grunewald near Berlin, his father told him: "Today I have made a discovery as important as that of Newton." ([27], p. 170). Even though Erwin would only have been seven years old at that time, this recollection should not be discounted. It seems, however, that Planck's earlier sense of the objective of science nagged at him. In spite of the importance of the quantum, the discovery was based on a mathematical expediency that even Boltzmann thought was unphysical.

Planck's formula for the radiation density was obviously correct, as the finest available data had shown. The mathematical formula was then readily accepted as true. The mathematical fact that Planck's formula was based on discrete elements of energy ε was, however, scarcely noticed. Planck himself had difficulty shedding the notion that this energy element was simply a mathematical hypothesis and not a real property of the energy distribution. Kragh points out that Planck's discomfort with the quantum is evidenced by his silence on the issue between 1901 and 1906 [160].

By introducing the quantum, Planck obtained the form of the complexions which then yielded the entropy (2.41) as a function of the resonator energy, and all the rest followed from this. The general form of the entropy from Wien's formula (2.43) then fit like a key to provide the identity $\varepsilon = h\nu$. But there were some serious problems. The critical relationship between radiation energy density and resonator energy density (2.5) was a result from classical physics, since the resonator obeyed Newtonian mechanics and the radiation was described in terms of Maxwell's equations.

Although Planck initially had difficulty accepting the idea that entropy could actually decrease at a particular instant, by 1912 he completely accepted this to be valid, along with the statistical nature of entropy [160].

The path was not straightforward, however, and there were issues along the way, which may have remained. Planck seems to have differed fundamentally from Boltzmann in his understanding of the statistical entropy in 1906. As we noted in the last chapter, Planck chose (formally) to deal with the energy and the entropy of individual resonators. The difficulty is that this was not simply a mathematical convenience. In 1906 Planck wrote of the disorder of the energy distribution and hence of the entropy of a single resonator. However, this is not actually consistent with Boltzmann's formulation, although at the Solvay Conference of 1911, he admitted that he went along with Boltzmann's definition of complexions ([64]; [227]; [20]).

By 1910 Planck had accepted that the quantum discontinuity was real and had to be taken into account at some point. He was convinced that Maxwell's electrodynamics completely described the radiation field. But the radiation field interacted with the resonators, which were quantized. So he chose the point at which it would do the

[1] After the unsuccessful assassination attempt on Hitler of 20 July 1944, Erwin Planck was arrested by the Gestapo. He was sentenced to death by the People's Court on 23 October 1944 and executed on 23 January 1945 in Berlin-Plötzensee.

least harm, which he decided was at the excitation of the resonator. Then in 1911 he reversed his position and decided that introducing the quantum in the emission process would raise the fewest problems. He seems to have been influenced by Hendrik Lorentz's (1853–1928) ideas on difficulties with the absorption of energy by a resonator in the high frequency regime. This picture in which the emission process was discontinuous, but the absorption process as well as the intrinsic energy were not, has been called Planck's second theory [64]. He presented it at the Solvay Conference in 1911 ([244], p. 156).

In 1912 in the foreword to the second edition of his book *Vorlesungen über die Theorie der Wärmestrahlungen* published in 1913, Planck wrote: "An actually new principle simply cannot be expressed in a functioning model based on old laws." Planck's great dilemma seems to have been this: how to escape from the bounds of the functioning model, while respecting the principles on which it is based [227].

3.3 Einstein and Photons

On 17 March 1905 Einstein sent a paper *Über einen die Erzeugung und Verwandlung des Lichtes betreffenden heuristischen Gesichtspunkt* (Regarding the emission and transformation of light from a heuristic point of view) to *Annalen der Physik* [96]. This is the paper often referred to as the photoelectric effect paper, although the photoelectric effect is only one of the examples considered. Einstein's objectives in the paper went well beyond the difficulties involved in understanding the photoelectric effect ([157], p. 20).

Einstein had made himself one of the world's experts in statistical mechanics during the years 1902–1904 by developing the theory that had been produced by Gibbs in 1902 [123]. In 1902 Einstein had only limited contact with the work of Boltzmann and was completely unaware of the treatise by Gibbs ([215], p. 55). Nevertheless, this effort provided Einstein with a deep understanding of the structure of matter and how the observable properties could be obtained from the atomic picture. He was also thoroughly familiar with the electrodynamics of Maxwell, as his paper on the special theory of relativity would demonstrate in June of 1905.

In the opening sentence of the paper he wrote that: "There is a profound and formal difference between the theoretical ideas that physicists have formed concerning gases, and other ponderable bodies, and Maxwell's theory of electromagnetic processes in so-called empty space." ([96], translation in [157], p. 21). And energy presents the same problem. The energy of ponderable matter is located in the atoms, molecules, and electrons, while the energy of the electromagnetic field is a continuous function of the space and time variables.

The wave picture of light had, of course, accurately described optical phenomena. However, Einstein pointed out that all optical observations involve time averages. He then suggested that there may be serious contradictions if we consider the emission and absorption of electromagnetic radiation. It seemed to him that such phenomena as the emission of blackbody radiation or the emission of cathode rays (electrons) by ultraviolet radiation might be much better understood if the energy in light were treated as discontinuous.

Considering that Maxwell's electromagnetic theory of light, supported by the luminiferous aether, was the crowning jewel of 19th century physics, Einstein was going against the very basis of physical theory at this point. He wrote that he would treat electromagnetic energy as discontinuous, consisting of a finite number of localized (*lokalisierten*) energy quanta which would move without division and become totally absorbed upon interaction with matter.

Then he announced that he would describe the thought process that led him to this conclusion in the hope that some researchers might find it helpful in their investigations [96]. Most of his readers did not follow him down the path. When he recommended Einstein for membership of the Prussian Academy and as director of the Kaiser Wilhelm Institut für Physik, Planck indicated that Einstein sometimes missed the mark, as in this paper, but that this should not be held against him ([157], p. 20, [215], p. 312).

In what follows we shall consider each section of Einstein's paper separately. This should provide a better understanding of the various points he made.

3.3.1 Regarding One of the Difficulties Encountered by the Theory of "Black Radiation"

Einstein first defined the system he wished to consider. He chose a hohlraum for which the internal walls of the cavity were totally reflecting. Within the cavity he assumed there were monatomic gas molecules[2] and free electrons. He assumed that all collisions would be elastic. In addition, he assumed that there were electrons located at widely separated points in the cavity wall and linked together by longitudinal forces. These wall electrons would interact elastically with the free electrons and atoms in the cavity and emit and absorb the blackbody radiation in the cavity according to the Maxwell theory of electromagnetism. He called these wall electrons *resonators*. Einstein's system was then the one considered by others such as Rayleigh and Planck.

The resonators in the cavity wall, the gas molecules and free electrons within the cavity, and the blackbody radiation in the cavity were all in thermodynamic equilibrium, and the thermodynamic temperature was uniform throughout. Einstein could then apply the well-known equipartition principle to the particles, including the resonators.

According to the equipartition principle ([138], pp. 175–177), each quadratic energy term for each particle would have an average energy of $(1/2)\,(R/N)\,T$, where

[2] Assuming a monatomic gas meant that only translational energy would have to be considered, which simplified calculations. However, Einstein would also have been aware of the difficulty in accounting for the vibration and rotational energies in molecules. There was wisdom in avoiding such issues.

R is the universal gas constant and N is what Einstein called the "actual number of molecules" in a gram-equivalent. This was, of course, Avogadro's number, which was only approximately known in 1905. Einstein could have written $k = R/N$ as the Boltzmann constant, but he chose not to. Each resonator had quadratic kinetic and potential energy terms, giving each a time averaged energy of

$$\bar{E} = \left(\frac{R}{N}\right) T. \tag{3.1}$$

From the equipartition theorem, the average translational kinetic energy of each (monatomic) molecule is $(3/2)(R/N)T$. Therefore the time averaged energy of each resonator is $(2/3)$ that of each molecule. In the event that through any action—in this case through the radiation—the energy of a particular resonator is greater or less than the average, the resonator energy will be brought back to the average through collisions with the molecules and the free electrons. A dynamical equilibrium is then actually possible only when each resonator has the average energy (3.1).

The resonators are thus in equilibrium with the blackbody radiation, which will be present in the cavity, and the average energy of a resonator in the frequency range $\nu \to \nu + d\nu$ will be given by

$$\bar{E}_\nu = \frac{c^3}{8\pi\nu^2}\rho_\nu(T), \tag{3.2}$$

where Einstein wrote L rather than c for the speed of light. We have reverted to c to avoid confusion and we have used Einstein's notation ρ_ν for the radiation energy density. Equation (3.2) is Planck's relationship for the resonator energy density in terms of the radiation energy density. Einstein cited Planck's 1900 *Annalen der Physik* paper as his source for (3.2) [223].

Combining Eqs. (3.1) and (3.2) then yields the density (per frequency interval) of the radiation (field) as

$$\left(\frac{R}{N}\right) T = \bar{E} = \bar{E}_\nu = \frac{c^3}{8\pi\nu^2}\rho_\nu(T).$$

The radiation energy density is thus

$$\rho_\nu(T) = \frac{R}{N}\frac{8\pi\nu^2}{c^3}T. \tag{3.3}$$

Einstein then pointed out that (3.3) not only fails to agree with experiment, but also shows that we can no longer speak of an energy distribution between the radiation[3] and matter (the resonators). Specifically, as the frequency range of the resonators increases, the energy in the radiation field resulting from the resonators increases without limit. That is,

[3] Here Einstein uses the term "ether" rather than radiation.

$$\int_0^\infty \rho_\nu (T) \, d\nu = \frac{8\pi}{c^3} (R/N) T \int_0^\infty \nu^2 d\nu = \infty. \tag{3.4}$$

This is what Paul Ehrenfest (1880–1933) would later call the "*ultraviolet catastrophe*" ([148], p. 13). Rayleigh had obtained a similar result in his paper of 1900 (see 2.4) with $\lambda^{-4} = \nu^4/c^4$), but he had made no mention of the difficulty [173]. This, too, did not seem to concern those working in this area ([157], pp. 21–22). Einstein was the first to point out the real difficulty this raises. This non-physical result simply could not be tolerated.

3.3.2 Regarding the Planck Result for Elementary Quanta

In what follows, Einstein said he would show that the elementary quanta and the theory of blackbody radiation proposed by Planck were to a certain extent independent of one another.

As he pointed out, all previous experiments were in agreement with the Planck formula

$$\rho_\nu (T) = \frac{\alpha \nu^3}{\exp(\beta \nu / T) - 1}, \tag{3.5}$$

and provided numerical values for α and β, which he would use later in the paper. He cited [226]. In the long wavelength (low frequency) limit, Einstein noted that this gives

$$\rho_\nu (T) = \frac{\alpha}{\beta} \nu^2 T, \tag{3.6}$$

which agrees with experiment (i.e., the Rubens and Kurlbaum data). But then he went farther. He combined (3.3) and (3.6) to give

$$N = \frac{8\pi\beta}{c^3 \alpha} R = 6.17 \times 10^{23}, \tag{3.7}$$

and noted that the mass of an atom of hydrogen was then 1.62×10^{-24} g. The result in (3.7) is, of course, Avogadro's number (for a gram equivalent), as we noted above. This, Einstein wrote, agrees with the value that Planck had found by other means and also with the values that others had obtained.

Einstein concluded that for high energy densities and long wavelengths the analysis based on a combination of standard Maxwell theory with statistical thermodynamics (Wien's law and the equipartition principle) produces results that agree with experiment. He noted, however, that "for short wavelengths and low densities these fail completely."

So what should be done next? Klein indicates that Einstein's answer was to proceed *boldly* ([157], p. 23) Einstein wrote: "In what follows we shall consider blackbody radiation in terms of experimental results, ignoring any picture of the production or dispersion of the radiation."

3.3.3 *Regarding the Entropy of the Radiation*

Einstein acknowledged that some of the ideas he was going to present had already been developed in a famous work by Wien.

He accepted that if the radiation energy density $\rho(\nu)$ (Einstein now included a dependence on ν rather than just putting ν as a subscript) was known for all frequencies (i.e., (3.5)), then the entropy of the radiation field in the cavity of volume V would have the form

$$S = V \int_0^\infty \varphi(\rho, \nu)\, d\nu, \tag{3.8}$$

where φ is the *entropy density* of the radiation field per frequency interval $d\nu$. For the radiation to be "blackbody" and in equilibrium with the molecules, electrons, and resonators, S had to be a maximum, subject to the constraint of constant radiation energy. That is, using the method of *Lagrange undetermined multipliers*, with λ as Lagrange multiplier,

$$\delta \int_0^\infty [\varphi(\rho, \nu) - \lambda\rho]\, d\nu = \int_0^\infty \left[\frac{\partial}{\partial\rho}\varphi(\rho, \nu) - \lambda\right] \delta\rho\, d\nu = 0$$

(see, e.g. [138], pp. 369–370), for arbitrary variations $\delta\rho$ in the radiation density ρ. This requires that

$$\frac{\partial}{\partial\rho}\varphi(\rho, \nu) - \lambda = 0.$$

The Lagrange multiplier λ is independent of ν. Therefore $\partial\varphi/\partial\rho$ must also be independent of ν.

We then return to (3.8), taking the volume to be $V = 1$, and consider a reversible change in the system temperature dT. The change in temperature will result in a change in the energy density ρ and a corresponding change in the entropy of the radiation field. From (3.8), this change in the entropy is

$$\begin{aligned} dS &= \int_{\nu=0}^{\nu=\infty} \frac{\partial\varphi(\rho, \nu)}{\partial\rho} d\rho\, d\nu \\ &= \frac{\partial\varphi(\rho, \nu)}{\partial\rho} dE, \end{aligned} \tag{3.9}$$

since the partial derivative $\partial\varphi/\partial\rho$ is independent of ν. Because the energy change dE is reversible, thermodynamics requires the change in energy of the system at constant volume V to be

$$\mathrm{d}S = \frac{1}{T}\mathrm{d}E, \tag{3.10}$$

by the Gibbs equation ([138], p. 29). Then, comparing (3.10) with (3.9), we have

$$\frac{\partial \varphi}{\partial \rho} = \frac{1}{T}. \tag{3.11}$$

Equation (3.11) is what Einstein called the *law of blackbody radiation.*

3.3.4 Limiting Form of the Entropy for Low Density Monochromatic Radiation

The next step was to integrate (3.11) to obtain the entropy density φ, noting that $\varphi = 0$ when $\rho = 0$. To carry out the integration, Einstein needed an equation for T in terms of ρ. For this he elected to use Wien's law (2.3), which is, in terms of frequency,

$$\rho = \alpha \nu^3 \exp\left(-\beta \frac{\nu}{T}\right). \tag{3.12}$$

He noted specifically that Wien's Law was not applicable in the long wavelength region he had considered previously, but that for large values of ν/T, it was in complete agreement with experiment (see Fig. 2.9). He also noted that as a consequence his results would be applicable only in a certain region.

From Wien's law (2.3), we have

$$\frac{1}{T} = -\frac{1}{\beta\nu}\log\frac{\rho}{\alpha\nu^3}.$$

Then (3.11) is

$$\frac{\partial \varphi}{\partial \rho} = -\frac{1}{\beta\nu}\log\frac{\rho}{\alpha\nu^3}, \tag{3.13}$$

and, upon integrating (3.13),

$$\varphi(\rho, \nu) = -\frac{1}{\beta\nu}\rho\left[\log\frac{\rho}{\alpha\nu^3} - 1\right] \tag{3.14}$$

for the radiation occupying the volume V with frequency in the interval $\nu \to \nu + \mathrm{d}\nu$. The energy of the radiation in this interval is $E = V\rho\mathrm{d}\nu$ and the entropy of the radiation in this frequency interval is $S = V\varphi(\rho, \nu)\mathrm{d}\nu$. Hence,

$$S = V\varphi(\rho, \nu)\,\mathrm{d}\nu = -\frac{E}{\beta\nu}\left[\log\frac{E}{\alpha V\nu^3\mathrm{d}\nu} - 1\right]. \tag{3.15}$$

If the same radiation (same temperature and frequency interval) occupied a volume V_0, the entropy would be S_0 with V replaced by V_0 in (3.15). The change in entropy between these two conditions is

$$S - S_0 = \frac{E}{\beta \nu} \log\left(\frac{V}{V_0}\right), \tag{3.16}$$

which, as Einstein pointed out, is the same form as the entropy difference for an ideal gas or weak solution resulting from a reversible change in volume (at constant temperature). This could also be interpreted, he proposed, through the formulation of the entropy in terms of probabilities, as Boltzmann had done.

3.3.5 Molecular Theoretical Considerations Regarding the Dependence of the Entropy of Gases and Weak Solutions on Volume

In the case of molecules, Einstein pointed out, the word "probability" is often used in a context that is not covered by the concept of probability. In particular, the expression "situations of equal probability" is often used in situations for which the conditions are reasonably well known, but not precisely enough for an exact definition. Einstein stated that he would show in a later work that the use of "statistical probability" was sufficient in such thermal transformations. He hoped that he could at last put aside any logical problems regarding the use of the Boltzmann principle here. At least, he would be considering here a very general principle and a rather specific case. Naturally, Einstein used W (from the German *Wahrscheinlichkeit*) to denote the probability, and we shall stay with this notation.

If it makes sense to speak of the probability of the state of a system, and if every increase in entropy can be associated with a transition to a more probable state, then the entropy of a state is determined by the instantaneous probability of the state in which the system finds itself.[4] Then two non-interacting systems have entropies S_1 and S_2 which are functions of the corresponding probabilities

$$S_1 = \varphi_1 (W_1),$$

$$S_2 = \varphi_2 (W_2).$$

And if we consider these two systems as a single system with an entropy S and a state probability W, then

$$S = S_1 + S_2 = \varphi (W),$$

[4] These claims about statistical probabilities, and our ability to have instantaneous knowledge of such states, may not be trivial. But Einstein's contention that this reasoning is commonplace is correct.

where the combined probability W is given by

$$W = W_1 \cdot W_2 .$$

From these equations it follows that

$$\varphi\left(W_1 \cdot W_2\right) = \varphi_1\left(W_1\right) + \varphi_2\left(W_2\right) ,$$

and finally that

$$\begin{aligned} \varphi_1\left(W_1\right) &= C \ln\left(W_1\right) + \text{constant} , \\ \varphi_2\left(W_1\right) &= C \ln\left(W_1\right) + \text{constant} , \\ \varphi\left(W\right) &= C \ln\left(W\right) + \text{constant} , \end{aligned}$$

where C is a universal constant. From kinetic theory, we know that this constant is R/N. Then if S_0 is the entropy of a reference state and if W is the relative probability of the state with entropy S, we have

$$S - S_0 = \frac{R}{N} \log\left(W\right) . \tag{3.17}$$

At this point Einstein chose to consider a volume V_0 in which there is a number n of moving points (e.g., molecules) on which he chose to focus attention. In this same volume V_0, there is also another set of arbitrary moving points of another sort. He made no assumption regarding the laws of motion except to insist that no part of the volume or direction be exceptional. He chose the number of particles of the first sort be be small enough for the forces of interaction among them to be neglected.

When the n moving points, considered as an ideal gas or ideal solution, occupied the volume V_0, Einstein denoted the entropy of the n points by S_0. If they were then to occupy a volume V with no other change in the system, he denoted their entropy by S. He could now specify the entropy difference as deduced from the Boltzmann principle.

Einstein first asked for the probability that the n moving points were suddenly located in a volume V, when in the prior instant they were in the volume V_0. For this probability, he said, we can offer only the "statistical probability"

$$W = \left(\frac{V}{V_0}\right)^{\mathrm{n}} , \tag{3.18}$$

whence the Boltzmann principle gives the entropy difference as

$$S - S_0 = R\left(\frac{n}{N}\right) \log\left(\frac{V}{V_0}\right) . \tag{3.19}$$

According to Einstein, it was remarkable that the Boyle–Gay-Lussac law and the law of osmotic pressure could be so easily shown thermodynamically to result from equation (3.19), without placing any further requirement on the motion of the molecules. And he carried out the calculation that resulted in the ideal gas law in a footnote.

3.3.6 *Interpretation of the Expression for the Dependence of the Entropy of Monochromatic Radiation on Volume According to the Boltzmann Principle*

Einstein had already obtained an equation for the change in entropy of the (monochromatic) radiation in the cavity resulting from a change in volume, viz., (3.16). And since

$$\frac{1}{\beta\nu} \ln\left(\frac{V}{V_0}\right) = \frac{R}{N} \log\left(\frac{V}{V_0}\right)^{NE/R\beta\nu},$$

he could then write (3.16) as

$$S - S_0 = \frac{R}{N} \log\left[\left(\frac{V}{V_0}\right)^{NE/R\beta\nu}\right]. \tag{3.20}$$

If we then compare (3.20) with the Boltzmann principle in the form (3.17), we conclude that the probability that monochromatic radiation of frequency ν contained in a volume V_0 with reflecting walls at a particular instant of time becomes spontaneously contained within a volume V is

$$W = \left(\frac{V}{V_0}\right)^{NE/R\beta\nu}. \tag{3.21}$$

The energy quanta have magnitude $\varepsilon = R\beta\nu/N$. The exponential of the volume ratio (V/V_0) is then $E/(R\beta\nu/N) = E/\varepsilon = n$, which is the number of energy quanta present. Therefore, (3.21) is identical to (3.18). Using this result, Einstein concluded that low density monochromatic radiation (within the limits of Wien's radiation formula) behaves thermodynamically as if it consists of independent energy quanta of magnitude $R\beta\nu/N$ in the frequency interval $\nu \rightarrow \nu+d\nu$.

If we compare (3.12) with the Planck formula (2.47) in the high frequency region, we see that

$$\beta = \frac{h}{k}.$$

Then with $R/N = k$, the energy of the quantum is $\varepsilon = R\beta\nu/N = h\nu$, as Planck had discovered. Although Einstein denoted the energy quantum by $R\beta\nu/N$ throughout his paper, we shall use the modern $\varepsilon = h\nu$ in most of our discussion beyond this point.

We also note that Einstein made no proposal regarding the laws of motion of the radiation quanta. He was able to ignore the laws of motion because the density of blackbody radiation in the high frequency (short wavelength) region is low. This was appropriate for his use of Boltzmann's entropy formulation, which is based on statistical probabilities.

Einstein then sought the average energy of the "black radiation" quanta in order to compare this with the average translational energy of the molecules, which was $(3/2)\,(R/N)\,T$. Using Wien's formula (3.12) for the radiation energy density ρ, the average energy per quantum is[5]

$$\frac{\int_0^\infty \mathrm{d}\nu \alpha \nu^3 \exp\left(-\beta\nu/T\right)}{\int_0^\infty \mathrm{d}\nu \left(N/R\beta\nu\right) \alpha\nu^3 \exp\left(-\beta\nu/T\right)} = 3\left(\frac{R}{N}\right)T. \tag{3.22}$$

Now that he had shown that radiation, in sufficiently low density to be described by Wien's formula, could be considered to consist of particles in the sense of the dependence of the entropy of the radiation on volume, Einstein suggested we should ask whether the present laws of emission and transmission of light were in agreement with expectations in the case where the light consisted of particles.

3.3.7 Regarding Stokes Law

Einstein considered photoluminescence in which monochromatic light of frequency ν_1 (quanta of energy $h\nu_1$) is directed on matter and light with frequency ν_2 is emitted from the matter. Other, lower frequencies $\nu_3, \ldots, \nu_s$ may be emitted and the matter may heat up as well. The energy bound found by considering light to be made up of quanta is

$$h\nu_2 \leq h\nu_1,$$

or

$$\nu_2 \leq \nu_1.$$

This is Stokes' law. He also noted that every incoming light quantum is capable of producing the sort of elementary process that results in the emission of other light quanta. Therefore, there is no lower boundary for the intensity of the light directed on the surface below which there is no light produced.

[5] We note that the integral in the denominator contains the factor $(R\beta\nu/N)^{-1}$, which is the number of quanta per unit energy, and that the Wien formula has the units of energy density per frequency interval $\mathrm{d}\nu$. The result then has the units of energy per quantum.

3.3.8 *Regarding the Production of Cathode Rays by Illumination of Solid Surfaces*

Einstein then turned to the work by Lenard, dealing with the production of cathode rays by illumination of solid material [171]. This is the photoelectric effect. Einstein called this a "trailblazing" (*Bahnbrechenden*) work.

First Einstein pointed out that it is particularly difficult to understand the results of Lenard's experiments if one starts from the idea that the energy of light is continuous. Then he considered the picture obtained if one assumes that the light impinging on the surface is quantized, as he was proposing. There are many possibilities for what may happen within the body. But the simplest was that the light quantum is entirely given up to a single electron (cathode ray). Although in 1905 a detailed picture of the atom had already been developed by Thomson (see Sect. 1.6), Einstein ruled out any complications due to possible ionization of molecules. He simply considered that the electron would give up some energy before reaching the surface of the material body. To leave the surface the electron also had to overcome the potential barrier known as the work function, which he denoted by P. Energy conservation then yielded the kinetic energy of the electron leaving the material body as

$$h\nu - P\,. \tag{3.23}$$

This is the maximum kinetic energy because the quantum of energy may be distributed among a number of electrons. If a positive potential Π (Einstein's notation) is applied to the body, then no electrons will leave the body if

$$\Pi q_e = h\nu - P, \tag{3.24}$$

where q_e is the electron charge, or

$$\Pi Q_e = Nh\nu - P',$$

where Q_e is the charge on a gram equivalent of a singly charged ion and $P' = NP$. Einstein then chose ν to be the ultraviolet bound in sunlight (1.03×10^{15} s^{-1}), set $P = 0$, presumably because no reasonable value was known, used the known value for β from Wien's law, and obtained 4.3 V for the stopping potential, which was in basic agreement with Lenard's results. Einstein wrote: "As far as I can see, the results of our proposal are not in disagreement with the experimental results of Lenard."

He also noted: "If the formula we have derived is correct, Π must be a linear function of the frequency ν of the exciting light. Represented in Cartesian coordinates, Π versus ν would then be a straight line, the slope of which is independent of the material being used in the experiment." This experiment had not yet been carried out in 1905.

3.3.9 Regarding the Ionization of a Gas by Ultraviolet Light

Einstein ended the paper with a remark on the ionization of gas molecules and compared the predictions of his theory with the data of Johannes Stark (1874–1957).

3.4 Summary

In this chapter we have looked at the responses of two of the key figures who contributed to and responded to the quantum in radically different ways.

Planck's concern had shifted from thermodynamics to Maxwellian electrodynamics. It seems that he had become comfortable with the statistical basis of classical thermodynamics. But he still found it hard to move from there to the idea that the sources of the electromagnetic waves might absorb and emit in quantum jumps.

Einstein was much younger than Planck. He was also a very different kind of person. The paper we have analyzed here was of primary importance in the development of thought on the quantum theory. It was the first of four papers written by Einstein in 1905, in what is often referred to as his *annus mirabilis* (miracle year), although that fact is incidental to our purpose.

Chapter 4
Quantum Atoms

> If you want to be a physicist you must do three things – first study mathematics, second, study more mathematics, and third, do the same.
>
> Arnold Sommerfeld

4.1 Introduction

In this chapter we shall look at Bohr's atomic model and the extension of Sommerfeld's.

We will follow Bohr's path from Copenhagen to Cambridge and then to Manchester. When Rutherford's group at Manchester identified the nucleus, attempts to model the atom faced an entirely new problem. A finite number of electrons had to be arranged dynamically around the tiny positive nucleus. But Maxwellian electrodynamics required that electrons in orbits around the nucleus radiate away energy in the form of electromagnetic waves.

As his postdoctoral year in England was coming to an end, Bohr presented a proposal to Rutherford for an atomic model. We shall see how this led him to his model, after coming across Johann Balmer's (1825–1898) numerical series for the hydrogen spectrum. We shall follow Bohr's thoughts as he expressed them in the 1913 paper, where he presented his model [15].

Then we shall turn to Sommerfeld, who contacted Bohr after reading his 1913 paper. We shall look at Sommerfeld's role in some detail, because he was in fact a visionary teacher. Just like another great teacher before him, Jacobi, he brought the results of his research into his classes as soon as he could, and he worked to build a strong connection between theory and experiment.

Sommerfeld mentioned to Bohr that he hoped to undertake a study of the Zeeman effect, something Bohr already realized would be difficult. But Sommerfeld had

C. S. Helrich, *The Quantum Theory—Origins and Ideas*, History of Physics,
https://doi.org/10.1007/978-3-030-79268-8_4

found his own path and was able to put together a formulation that provided a way forward. However, this was not yet a real quantum theory and Sommerfeld finally gave up modeling the atom and moved into something he referred to as "number mysteries."

4.2 Niels Bohr

4.2.1 Doctorate and Cambridge

Bohr was born and educated in Copenhagen, Denmark. His father was a professor of physiology at the University of Copenhagen and his mother was from a prominent Jewish family in Copenhagen. Bohr's centrality in the new physics can be attributed in part to his personality, but also to the fact that he came from a small country which acknowledged the importance of science, but was not seen as a scientific powerhouse. Bohr's intellectual inclinations were certainly nurtured by the intellectual atmosphere at home (cf. [238]).

Bohr received his Ph.D. from the University of Copenhagen in May of 1911 with a dissertation on electrons in metals, using the model developed by the German theoretical physicist Paul Drude (1863–1906). In Drude's model, solid structures resulted from an ordering of their constituent ions, i.e., atoms minus their valence electrons. He assumed that the valence electrons moved freely like a gas within the solid. Consistent with the gas model, the collisions of the electrons with the ions and among themselves were taken to be almost instantaneous, despite the long range of the Coulomb force (cf. [22]; [89]). This model was very successful and still forms the basis for simple pictures and for obtaining rough estimates of the properties of conductors ([7], Chap. 1).

By 1911 physicists were already realizing the limitations of the Drude theory, and Bohr discovered that, if the electrons were treated with the methods of statistical mechanics, Drude's theory could not explain the magnetic properties of metals. This result in particular led Bohr to suspect that a non-mechanical description would be necessary for the treatment of electrons in metals. And this left an indelible mark on Bohr's approach to physics and his construction of the atom, which is our topic here ([129], p. 66).

After his Ph.D., Bohr took a postdoctoral position with Thomson at the Cavendish Laboratory in Cambridge. This was not a happy situation for Bohr. His English was poor, which did not help attempts at communication with Thomson. At one point he went to Manchester to take a short course in radioisotopes with Rutherford, who Bohr recognized was the growing figure in English physics. At one point the laboratory work was held up, and Bohr spent the time reflecting on the new demands regarding the nucleus and the limited number of electrons in the atom. Within a short time, he set aside his experimental work and spent all his time on model building ([129], p. 67).

4.2.2 Bohr's Model

In July of 1912 Bohr presented a memo to Rutherford for a discussion. The memo contained the points:

1. The nuclear atom is inherently unstable mechanically (Newtonian mechanics & Maxwellian electrodynamics).
2. Bohr proposed stability by fiat. The kinetic energy T of the electron in orbit would be given by a relation of the form

$$T = K\nu', \tag{4.1}$$

   where ν' is the frequency of orbital rotation.
3. The usual force balance will otherwise hold for the orbits.
4. Deformation of the orbit is prohibited.
5. Bohr assumed that the constant K was of the order of Planck's constant, h. He hoped to deduce this in some way.
6. Bohr's force balance was simply

$$\frac{m_{\mathrm{e}}v^2}{a} = \frac{2T}{a} = \text{Coulomb force} = \frac{eE}{a^2}, \tag{4.2}$$

   where a is the orbit radius and E is the charge on the nucleus ([129], p. 68) The kinetic energy is given by (4.1).

We shall use Gaussian units here, like Bohr.

The winter of 1912 found Bohr back in Copenhagen with a teaching job at the university. In Copenhagen he read John W. Nicholson's (1881–1955) papers, in which he proposed his own atomic model. Nicholson had developed a nuclear model independently of Rutherford, with electrons rotating about the nucleus. He had suggested that the spectrum of an atom arises from perturbations in the electron orbital motion perpendicular to the plane of the orbit, which he had shown were stable, although his analysis considered only atoms with up to five electrons.

In Nicholson's model, the perturbed frequencies were multiples of the orbital frequency ν', which could be chosen arbitrarily. For a particular choice of ν', the results for the emitted frequencies agreed with the spectrum of the hydrogen atom as observed from certain nebulae. And ionized forms of Nicholson's atoms gave other lines in the solar corona.

Bohr noted that the angular momenta in Nicholson's atom appeared in multiples of $(h/2\pi)$. This quantization of the angular momentum of the electron struck him as important ([129], p. 69).

Bohr was bothered by the fact that Nicholson's atom provided the apparently correct emission spectrum for hydrogen, while his own model provided no emission at all. He also harbored a concern that was common to everyone trying to model atomic structure. The spectra of atoms were simply considered too complicated to be used as the basis for any model. We have seen this in the fact that Thomson made no attempt to calculate spectra from his elaborate atomic model, although he had sufficient electrons to provide some kind of scheme ([129], p. 69).

According to Heilbron, Bohr first confronted the emission spectrum of hydrogen in a conversation with a colleague in March of 1913. This colleague had asked Bohr how his work on the atom was going and in particular whether his model had anything to say about the spectrum of hydrogen. Bohr admitted that his model was silent on the spectrum, although he may have made some remarks about how complex spectra were generally. His colleague then suggested that he should look at the analysis of the hydrogen spectrum by Balmer ([129], p. 70).

Balmer was a Swiss mathematician with a Ph.D. from the University of Basel. He remained in Basel, where he taught at a girls' secondary school and lectured at the university.[1] In 1885 he published an empirical formula for the wavelengths of the lines in the spectrum of hydrogen:

$$\lambda = \frac{h_B m^2}{m^2 - n^2},$$

where h_B is an empirical constant, m is an integer equal to 3, 4, 5 . . ., and $n = 2$. If we write this in terms of the frequency $\nu = c/\lambda$, then

$$\nu = R_H \left(\frac{1}{n_1^2} - \frac{1}{n_2^2} \right),$$

where $n_1 = n$ and $n_2 = m$ and R_H is the Rydberg constant for hydrogen ([33]; [148], pp. 67–68; [245], p. 366; [143], pp. 432–434).

Janne (Johannes) Rydberg (1854–1919) was a Swedish mathematician and physicist who had a rather disappointing career at Lund University, where he received his doctorate in 1879. The disappointment bore no relation to Rydberg's ability, and he was rated highly on the continent. Swedish physics considered the gathering of one's own data and subsequent analysis to be of the greatest importance, while Rydberg's mathematical work on spectra was based on data gathered by others. We may note that Rydberg's work preceded Balmer's ([242], [178]).

Bohr later said: "As soon as I saw Balmer's formula the whole thing was immediately clear to me." ([129], p. 70).

Bohr sent a draft of his first paper on atomic theory to Rutherford on 6 March 1913, and Bohr and Rutherford sat together to discuss it in July of 1913 ([238], p. 83; [2]). In this first of three papers on his model, Bohr began by describing Rutherford's nuclear atom and contrasted it with J.J. Thomson's model, in which the positive charge occupied the entire atomic volume. He admitted that this made it impossible to accept a balancing process of the kind Thomson had used. He also mentioned Planck's theory of the quanta and indicated that he would develop a theory based on the quantum [15].

[1] For a teacher in a Swiss/German secondary school to hold a Ph.D. is not unusual. This is a credit to the educational system and should not be considered a failure of the individual.

Consistent with his 1912 memo to Rutherford, Bohr considered an electron of charge e rotating around a nucleus of charge E. He admitted that according to classical electrodynamics the electron would emit radiation and spiral into the nucleus. Generally, an orbit would be elliptical. However, he suggested that there might be certain circular orbits that would be stable in the sense that an electron in one of these orbits would not emit radiation. In such a stable orbit of radius a, classical mechanics would hold.

The potential energy of the orbiting electron is

$$\Phi = -\frac{Ee}{a}. \tag{4.3}$$

The kinetic energy T of the electron in the stable orbit may be obtained from a radial force balance for the electron, viz.,

$$m_e \frac{v^2}{a} = \frac{Ee}{a^2}. \tag{4.4}$$

Then T is

$$T = \frac{1}{2}\frac{Ee}{a}. \tag{4.5}$$

From (4.3) and (4.5), the total orbital energy is

$$W = \frac{eE}{2a} - \frac{eE}{a} = -\frac{eE}{2a}, \tag{4.6}$$

where we have introduced Bohr's term W for this total orbital energy. We also notice that the total orbital energy is equal in magnitude to the kinetic energy. The negative sign indicates a bound orbit. Our equation (4.6) is one of the two equations labeled as (1) in Bohr's 1913 paper [15].

Bohr used ω to denote the rotational frequency of the electron in a stable orbit. This is what he denoted by ν' in his 1912 memo to Rutherford. The velocity of the electron in orbit is then

$$v = 2\pi a\omega\,,$$

and the total orbital energy, which is equal in magnitude to the kinetic energy of the orbiting electron, is

$$W = \frac{1}{2}m_e v^2 = \frac{1}{2}m_e\left(4\pi^2a^2\omega^2\right). \tag{4.7}$$

From (4.7), we have

$$\omega^2 = \frac{W}{2\pi^2 m_e a^2}. \tag{4.8}$$

With (4.6), Eq. (4.8) becomes

$$\omega^2 = \frac{2W^3}{\pi^2 m_e a^2} \frac{1}{e^2 E^2}, \tag{4.9}$$

or

$$\omega = \frac{\sqrt{2}W^{3/2}}{\pi \sqrt{m_e}} \frac{1}{eE}. \tag{4.10}$$

Equation (4.10) was the second of the two equations Bohr labeled as (1). [15] We note that W determines both the orbital radius a and the orbital frequency ω.

At this point Bohr noted that Planck quantized the radiation from electron oscillators as $\tau h\nu$, where τ was an integer. Bohr assumed that homogeneous radiation would be emitted at a frequency ν, which was half the orbital frequency ω, and that the amount of radiation emitted would be $\tau h\nu$. Then W, which is the magnitude of the total energy, is given by

$$W = \tau h \frac{\omega}{2}. \tag{4.11}$$

Equation (4.11) was labeled as (2) in Bohr's paper. [15]

Using (4.11) for the value of ω in (4.10), we obtain

$$W = \frac{2\pi^2 m_e e^2 E^2}{\tau^2 h^2}. \tag{4.12}$$

Then, eliminating W between (4.11) and (4.12), we obtain

$$\omega = \frac{4\pi^2 m_e e^2 E^2}{\tau^3 h^3}. \tag{4.13}$$

And with (4.12), Eq. (4.6) becomes

$$2a = \frac{\tau^2 h^2}{2\pi^2 m_e e E}. \tag{4.14}$$

The last three Eqs. (4.12), (4.13), and (4.14) were labeled as (3) in Bohr's paper. [15] With the (then) known values for e, e/m, and h, Bohr computed the electron orbital diameter $2a$ to be 1.1×10^{-8} cm, the orbital frequency ω to be $6.2 \times 10^{15}\ \mathrm{s}^{-1}$, and the binding (ionization) energy per charge W/e to be 13 V.

Then, in Sect. 2 of the paper, Bohr considered the hydrogen atom by choosing $E = e$. [15] The state τ thus has the orbital energy

$$W_\tau = \frac{2\pi^2 m_e e^4}{h^2} \frac{1}{\tau^2}. \tag{4.15}$$

For two such states τ_1 and τ_2 with energies W_{τ_1} and W_{τ_2}, the energy change in the transition between these states is

$$W_{\tau_2} - W_{\tau_1} = \frac{2\pi^2 m_e e^4}{h^2} \left(\frac{1}{\tau_2^2} - \frac{1}{\tau_1^2} \right). \tag{4.16}$$

In this transition, homogeneous radiation is emitted at frequency ν according to

$$W_{\tau_2} - W_{\tau_1} = h\nu. \tag{4.17}$$

That is,

$$\nu = \frac{2\pi^2 m_e e^4}{h^3} \left(\frac{1}{\tau_2^2} - \frac{1}{\tau_1^2} \right). \tag{4.18}$$

In this way, Bohr's model predicted a value of the Rydberg constant for hydrogen as $R_H = 2\pi^2 m_e e^4 / h^3 = 3.1 \times 10^{15}$ s^{-1}. In 1913 the accepted value of this constant was 3.290×10^{15} s^{-1}. That is Bohr's model had fit the hydrogen spectrum to within 5.8%. He noted this in Sect. 2 of his paper [15].

In Sect. 3 Bohr discussed his use of Eq. (4.11) in the way he imagined the atom of hydrogen was constructed from a free nucleus and a free electron. One can produce the Bohr atom without considering the formation of the atom, and this is normally done in elementary textbooks if the derivation is provided at all. Nevertheless, Bohr did consider a general release of energy in forming the atom $W = f(\tau) h\omega$ and concluded that the final form of the spectrum would still be given by (4.18). [15]

4.3 Arnold Sommerfeld

Königsberg was Sommerfeld's home town, so enrolling at the Albertus University in Königsberg after his Gymnasium Abitur (preparatory high school graduation) may have been logical enough. However, remaining solely at one university was unusual in Germany. He also became part of a fraternity (*Burschenschaft Germania*), which obliged his participation in drinking bouts and fencing duels. One of these left a visible scar on Sommerfeld's forehead. In the summer of 1891, he conceived and wrote his dissertation and received his doctorate in mathematics at Königsberg under Ferdinand Lindemann (1852–1939) [110]. Figure 4.1 is a color postcard showing the Albertus University in Königsberg and the monument to King Friedrich Wilhelm III (1770–1840) as Sommerfeld would have known them.

David Hilbert (1862–1943) was *Privatdozent* at Königsberg while Sommerfeld was a student. Through the influence of Felix Klein (1849–1925), Hilbert was given a post as professor of mathematics at Göttingen in 1895. Under the leadership of Klein and Hilbert, Göttingen became a world center for mathematics ([262], p. 342).

Through some personal connections, Sommerfeld managed to obtain a position for a year at the Mineralogical Institute in Göttingen, after which he spent two years as assistant to Klein. Part of his assignment was to copy Klein's lecture notes in legible form for the reading room. He wrote his *Habilitationsschift* there on diffraction and stayed at Göttingen as *Privatdozent* in mathematics [110].

Fig. 4.1 Color postcard showing the Albertus University in Königsberg and the monument to King Friedrich Wilhelm III (Public Domain)

After Göttingen, Sommerfeld accepted an appointment at the Bergakadamie (mining school) in Clausthal in 1897, which provided him with the financial support to marry. Then in 1900, as a result of Klein's efforts, he became a professor of engineering at the RWTH in Aachen (see Sect. 2.3 Fig. 2.8).

In 1906, after 6 years at Aachen, Sommerfeld was given a position as professor of theoretical physics and director of the new Institute of Theoretical Physics at the Ludwig Maximilian University in Munich (LMU) ([92], [110]). Figure 4.2 is a modern photograph (2005) of the entrance to the main building of the LMU

4.3.1 Thoughts and Ideas

In the years prior to his arrival in Munich, Sommerfeld's interest and ability in teaching were evident. Just as a former professor of mathematics, Jacobi, had once done at Königsberg, Sommerfeld brought the results of his research to his lectures ([244], p. 14) He also built his lectures around the problems facing physics and engineering to bring students an understanding of how to face real problems ([244], pp. 2, 3). And he engaged students in discussions after evening meals at his home or as part of strenuous outings in the Bavarian Alps ([110]; [244], pp. 62, 64).

Fig. 4.2 Entrance to the main building of the Ludwig Maximilian University in Munich (2005) (Author: Seeott)

In 1922 Einstein wrote to Sommerfeld: "What I especially admire about you, is the way, at a stamp of your foot, a great number of talented young theorists spring up out of the ground." ([103]; [244], p. 48). The number of truly great physicists that came out of LMU while Sommerfeld was there indicates that his vision did indeed bring results. [110]

What Sommerfeld had formed at the LMU was essentially a school, in the sense of a locally defined group under the influence of a charismatic teacher. However, there was no common way of thinking in Munich, which might be taken as another characteristic of a school. Sommerfeld aimed in particular to develop independent thinking [92]. Today, however, there is the *Arnold Sommerfeld Center for Theoretical Physics* at LMU.

Sommerfeld insisted on having experimental facilities at the Institute of Theoretical Physics, an idea he may have gained in Aachen. These were the facilities used clandestinely by Sommerfeld's assistant Walter Friedrich to discover X-ray diffraction in 1912 (see Sect. 1.8). His attempts to mix both aspects of physics was, however, not as successful as he had hoped and it finally came to an end. [93]

At the *Naturforscherversammlung* of 1907, Sommerfeld publicly defended Einstein's special relativity. This placed him among the earliest converts along with Planck. Then at the *Naturforscherversammlung* of 1909, Sommerfeld met Einstein and they formed an immediate bond, in spite of their differences. The topic of interest to them at this first meeting was, however, not relativity but quantum theory. Sommerfeld was not prepared to follow Einstein here. He was more conservative regarding Maxwell's theory and chose to proceed more cautiously, as Planck had [110].

The following year, Sommerfeld's work and interactions with students brought about a change in his intellectual position. He became convinced of the importance of the quantum and went to Zurich to spend a week with Einstein, discussing light and the quantum theory with some excursions into relativity. Sommerfeld was moving toward introducing a formal ansatz in the classical (Maxwell/Lorentz) approach to the interaction of radiation with atoms. The action was evidently of central importance. So it seemed reasonable to make the Ansatz that, in any physical process such as the emission of radiation, including bremsstrahlung, or the photoelectric effect, the total action integral should be ([244], p. 150)

$$\int \mathcal{E}\mathrm{d}t = n\left(\frac{h}{2\pi}\right), \tag{4.19}$$

where $\mathcal{E}$ is energy and t is time. That is, it is the action that is quantized, rather than the energy. This idea was central in the paper he presented at the Solvay Conference of 1911.

The topic of the Solvay Conference of 1911 was "Radiation Theory and Quanta." Sommerfeld's name did not appear on the first communications regarding the conference, for reasons we have seen here. But the flurry of papers he sent in after his discussions with Einstein soon changed that. And Sommerfeld's paper at the conference drew considerable interest ([244], p. 149). However, the idea of quantization of the action finally led nowhere and Sommerfeld abandoned it in 1913. Nevertheless, the idea had been important to Nicholson and Bohr (see Sect. 4.2). This was the source of the quantization of the orbital angular momentum in the models of the hydrogen atom ([209]; [210]).

4.3.2 *Zeeman Effect*

In the summer of 1913, after Bohr's publication of his model of the hydrogen atom, Sommerfeld sent a very friendly personal communication to Bohr. Sommerfeld was becoming skeptical of atomic models, and fully admitted it. However, the accuracy of Bohr's calculation of the Rydberg constant had impressed him (see Sect. 4.2). Sommerfeld closed his letter by mentioning that he would like to investigate the Zeeman effect (the splitting of a spectral line into components by the application of a magnetic field) on the basis of Bohr's model. Through his own experience, Bohr knew this was not going to be simple [110].

The first observations of the effect of an applied magnetic field on radiation emitted by an element were reported by Pieter Zeeman (1865–1943) in 1896 [292]. These were published in English in the *Philosophical Magazine* in 1897 [293]. In 1902 Zeeman shared the Nobel Prize in Physics with Lorentz for his discovery of the *Zeeman effect*.

Zeeman's main measurements were of the broadening of the sodium D lines under the influence of a strong magnetic field. He discussed his results with Lorentz, who responded with an analysis of the motion of an electron orbiting in a central force field in the horizontal plane when acted upon by a magnetic field of intensity H in the vertical direction. The Newtonian equations describing the electron motion in this situation are

$$
\begin{aligned}
m_{\mathrm{e}}\frac{\mathrm{d}^2x}{\mathrm{d}t^2} &= -k^2x + q_{\mathrm{e}}H\frac{\mathrm{d}y}{\mathrm{d}t}\,,\\
m_{\mathrm{e}}\frac{\mathrm{d}^2y}{\mathrm{d}t^2} &= -k^2y - q_{\mathrm{e}}H\frac{\mathrm{d}x}{\mathrm{d}t}\,,
\end{aligned}
\tag{4.20}
$$

where m_{e} and q_{e} are the mass and charge of the electron. The solutions are periodic. If $H = 0$, the period is $T = 2\pi\sqrt{m_{\mathrm{e}}}/k$. If $H \neq 0$, the period becomes $T' = 2\pi\sqrt{m_{\mathrm{e}}}/k\left(1 \pm q_{\mathrm{e}}H/2k\sqrt{m_{\mathrm{e}}}\right)$. That is, Lorentz's electron theory predicted a separation of the line (into two lines). Zeeman only reported a distinct broadening. He was unable to discern a separation [293].

The physical situation Lorentz described is what became known as the normal Zeeman effect. There is another spectroscopic pattern that became known as the anomalous Zeeman effect, which is actually more common than the normal Zeeman effect. [118] Wolfgang Pauli (1900–1958) remarked that the anomalous Zeeman effect was both very beautiful and at the same time hardly understandable. He recalled an occasion when he was strolling rather aimlessly in the beautiful streets of Copenhagen when a colleague encountered him and asked him why he looked so unhappy. Pauli said that his response was rather fierce: "How can one look happy when one is thinking about the anomalous Zeeman effect?" [117]

Sommerfeld realized that to undertake a study of the Zeeman effect based on a Bohr model he needed first to generalize the quantum description to apply to systems with more than one degree of freedom, which would be difficult. Nevertheless he began to get promising results and in the winter semester of 1914–15 he was already including his initial results in his lectures.

He had expanded his action integral (4.19) by defining the *action variables* J_ϑ and J_{r} of analytical mechanics, which are, for the orbit with variables (r, ϑ) ([140], p. 190),

$$
J_\vartheta = \oint p_\vartheta \mathrm{d}\vartheta \quad \text{and} \quad J_{\mathrm{r}} = \oint p_{\mathrm{r}}\mathrm{d}r,
\tag{4.21}
$$

since $J_\vartheta = kh$ and $J_{\mathrm{r}} = n'h$. From these definitions he was able to obtain the Balmer formula for the energy of the nth hydrogen level as

$$
W = \frac{R_{\mathrm{H}}h}{n^2},
\tag{4.22}
$$

where $n = n' + k$. We note that (4.22) is (4.15) with $\tau = n$ ([129], p. 79).

For each of the Bohr quantum numbers n, Sommerfeld had a group of elliptical orbits identified by $k = 1, 2, 3, \ldots, n$. The orbit with $k = n$ was circular and the

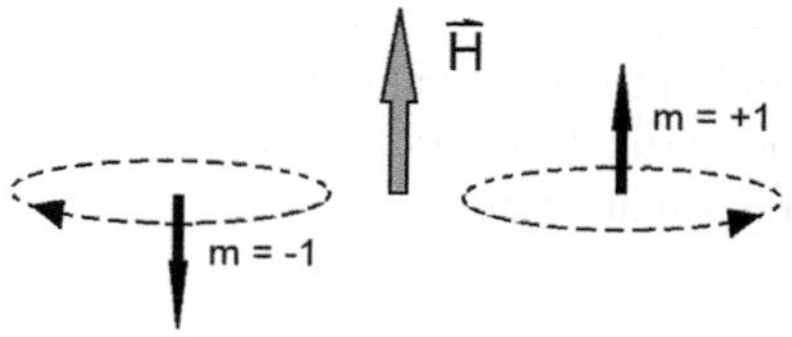

Fig. 4.3 Bohr–Sommerfeld orbit with $m = \pm 1$ in the presence of an external magnetic field of intensity **H**. The previously single level is split by $\pm \mu_B H$

one with $k = 1$ had the greatest eccentricity[2] ([129], p. 80). The quantum number n identifies a Bohr orbit and is now called the *principal quantum number*. The angle ϑ is the azimuthal angle and the quantum number k is the *azimuthal quantum number*. Finally, the *projection quantum number* m took on values $m = -k, -k+1, \ldots, +k$, with $m = 0$ excluded. The value of m specified the orientation of the orbit in space. For each positive value of m, there was a negative value of m. In the absence of a magnetic field, for example, all orientations were identical [117].

André Ampère (1775–1836) had suggested that magnetism in matter came from microscopic current loops. In the Bohr–Sommerfeld model, the orbiting electrons apparently fit Ampère's model. There was a magnetic moment associated with each elliptical orbit of magnitude $\mu_B m$, where $m_B = q_e (h/2\pi) / (2m_e)$ is the Bohr magneton, which was identified by the set of quantum numbers (n, k, m). The magnetic moment associated with that orbit is then $\boldsymbol{\mu}(n, k, m)$. In the presence of an external magnetic field of intensity **H**, this orbit has an additional energy $\boldsymbol{\mu}(n, k, m) \cdot \mathbf{H}$. The values $\pm m$ then implied a difference in the energy of the states resulting from the application of an external magnetic field, as Zeeman had observed.

Figure 4.3 is a simplified picture of the situation for the quantum numbers $n = 1$, $k = 1$, $m = \pm 1$, in the presence of an external magnetic field **H**.[3] The energies of the levels identified by the projection quantum numbers $\pm m$ are shifted by the amounts $\pm \mu_B H$. This shift in energies should be observable spectroscopically, as Zeeman had done. However, there was a difficulty associated with the polarization of the imposed wave in absorption measurements. If one treats the orbit classically, as Lorentz did, and as one had to in 1915, the effect of an electromagnetic wave on the orbits in Fig. 4.3 will be different depending on whether the wave is polarized along or parallel to **H**. Specifically the index of refraction of the matter being studied will depend on the polarization of the measurement wave. This phenomenon is known as birefringence. In this case the phenomenon would be magnetically induced birefringence. [117], [118]

In the spring of 1916 Sommerfeld had what seemed to be a definitive quantum theory for the Zeeman effect, and his student Paul S. Epstein (1883–1966) had similar success with the Stark effect (the splitting of a spectral line into components by the application of an electric field). The work by Sommerfeld and Epstein had a very positive effect on the acceptance of Bohr's model. And Sommerfeld wrote this up in

[2] If $k = n$ then $n' = 0$, which requires $J_r = 0$ and $p_r = 0$. This is a circular orbit.

[3] The state $n = 1$ is the ground state of hydrogen or the ground state of a single electron moving in a shielded potential with net charge $Z = +1$.

his *Atombau und Spektrallinien* (1919), which was often referred to as the "Bible" of atomic physics [110].

At this point Sommerfeld began a very fruitful working relation with the great experimentalist Paschen. Together they discovered a critical problem. Sommerfeld's multiple orbit scheme provided more orbits than Paschen had found experimentally in the fine structure of He^{+}. [219] Sommerfeld was able to make amends for this and bring his theory into line with Paschen's results by defining a set of selection rules, which were simple, but quite arbitrary ([129], p. 81).

Although the energy was dependent only on the principal quantum number for the nonrelativistic case, Sommerfeld showed that the variation in electron velocity in an elliptic orbit could contribute to a relativistic change in mass. He found that the resulting relativistic contribution gave good quantitative agreement with the observed spectral fine structure, which was considered a triumph for both the quantum and relativity theories [117].

In 1913 Peter Debye (1884–1966) published what we may consider to be a more rationalized expression for the quantization of the orbital angular momentum condition than Bohr had provided. Debye's condition was based on the action variable and was essentially the same as the first Sommerfeld condition in (4.21), i.e., $\oint pdq = nh$ in which q and p are the generalized coordinate and momentum of analytical mechanics. [75] This was dated 10 February 1913 (Utrecht) and seems to have preceded Debye's access to Bohr's publication. Debye then took up X-ray diffraction, hoping to obtain results that would provide evidence for the Bohr orbits, working with his young Swiss assistant Paul Scherrer (1890–1969). The fact that Debye was Dutch and Scherrer Swiss meant that their work was unaffected by the beginning of the war in 1914. However, they found no X-ray evidence to substantiate Bohr's ideas ([80], p. 189)

While at Göttingen, after 1913, Debye produced a treatment of the Bohr orbits, based on the Hamiltonian, which was more systematic than Sommerfeld's. Bohr used this in his paper for the *Proceedings of the Copenhagen Academy* in 1918 ([80], p. 190; [78]; [79]).

Related to this, Debye also produced a study of the Zeeman effect based on the orientations of the orbits ([76], [77]) This was published almost simultaneously with the publication by Sommerfeld which we considered above. In 1918 Bohr wrote: "Subsequently, Sommerfeld himself and Debye have on the same lines indicated an interpretation of the effect of a magnetic field on the hydrogen spectrum which [...] undoubtedly represents an important step towards a detailed understanding of this phenomenon." ([80], footnote on p. 190).

In the hydrogen atom, there is only one electron, so either one or the other of the two orbits we have drawn in Fig. 4.3 will be occupied. Hence, a single hydrogen atom in an external magnetic field will have a magnetic moment aligned or anti-aligned with the external magnetic field. This will also be the case for an atom with a single valence electron, which may be considered to have the hydrogen levels associated with a shielded nucleus with charge equal to unity. In each of these cases we would then expect a gas of these atoms to be composed of a statistical distribution of atoms with magnetic moments aligned or anti-aligned with the external magnetic field. The projection quantum number m then determines the spatial orientation of the orbital magnetic moment of the electron. There is thus a spatial quantization that may be used to distinguish atoms.

4.3.3 *Beyond Models*

The years between 1919 and 1926 saw a change in Sommerfeld's approach. He no longer believed that multi-electron atoms could be approached in a general theoretical manner based on first principles. He turned instead to the data. Where he had previously spoken of numerical harmonies he now used the term "number mysteries".

He and his students were successful. And the approach was being taken up by other atomic scientists. But this was not the way to a deeper understanding of atomic structure. [110] For that, we must look to the group working with Born at Göttingen and to Erwin Schrödinger (1887–1961) working alone and discussing mathematics with Hermann Weyl (1885–1955).

4.4 Summary

Bohr's atomic model was a bold step into the unknown. It was not built on strong physical principles, but it resulted in the Balmer series for the spectrum, once Bohr saw the connection. It could also be extended to atoms with single valence electrons and filled inner shells. Bohr's model was a success and remained a central topic for quite a long time.

Bohr had introduced the concept of an atomic state. Transitions between these states resulted in the emission or absorption of electromagnetic radiation. There was no explanation for how this transition might occur, and this seemed to some to be a shortcoming. The issue of the atomic state revealed something of a new way of thinking about the physics of the atom.

Sommerfeld approached the problem of the Zeeman effect using sound principles from analytical mechanics, and indeed he had some success. The atomic structure that emerged may be referred to as the Bohr–Sommerfeld model. We noted that Debye proceeded along similar lines to Sommerfeld, staying close to analytical mechanics. This would be the general path followed in a general route to a real quantum mechanics. As we noted, this also impressed Bohr.

Chapter 5
Experimental Evidence

Erst die Theorie entscheidet darüber, was man beobachten kann.
The theory first decides what can be observed.
Albert Einstein

5.1 Introduction

So far we have encountered the concept of the quantum in certain basic forms. The first of these was in Planck's discovery from the analysis of blackbody radiation. The second was Einstein's proposal of the photon. And the third was in Bohr's picture of the electronic structure of the atom, designed to produce the emission spectrum. Planck, Einstein, and Bohr were theoretical physicists who were well aware of the radical nature of their ideas. What was needed was further experimental work. In his chapter we will provide an outline of some of the critical experiments that gave support to the ideas ventured by Einstein and Bohr. These experiments would also involve some difficulties for the proposed theories. Our treatment will be basically historical.

5.2 Bohr Model

5.2.1 *Ionized Helium*

In 1896 the Harvard astronomer Edward Charles Pickering (1846–1919) found a series of strong lines in certain star spectra which seemed to be attributable to ionized helium He^{+}. According to Bohr's atomic model, the spectral lines from He^{+} should have frequencies corresponding to a Rydberg constant four times as large as that for

C. S. Helrich, *The Quantum Theory—Origins and Ideas*, History of Physics,
https://doi.org/10.1007/978-3-030-79268-8_5

hydrogen. That was because the nuclear charge in helium is twice that of hydrogen and $R_{He^+} = 2\pi^2 m_e \left(2e^2\right)^2 / h^3$. But careful measurements by spectroscopists with tubes purged of hydrogen produced a ratio of 4.00163. The accuracy of spectral measurements in 1913 was sufficient to contend that this result indicated a failure of the Bohr model.

However, Bohr realized where the problem lay. In fact, he had considered that the ratio of the electron mass to the nuclear mass to be negligible in orbital calculations. Without this approximation, the classical mechanical picture is that of an electron with a *reduced mass* orbiting the nucleus. When Bohr used the reduced mass in his calculation of the Rydberg constants for the hydrogen and ionized helium lines, he obtained a ratio of 4.00160 [117, 129, p. 75].

5.2.2 Moseley

In 1910 Rutherford hired Henry Gwyn Jeffreys Moseley (1887–1915) to perform classroom demonstrations. Apparently, Rutherford saw something promising in him, although Moseley said that his head was full of cobwebs after graduating from Oxford and that he could hardly think of research. But he nevertheless managed to land a research fellowship, which he decided to use to investigate X-rays.

Together he and the theoretician Charles (Galton) Darwin, who had been working with Rutherford and Bohr on atomic theory, approached Rutherford with a proposal to begin X-ray studies on the structure of atoms. They received Rutherford's endorsement, but there was no one in Manchester with sufficient expertise in this area, so Moseley contacted William Henry Bragg at the University of Leeds [63, p. 320].

We may recall that X-ray diffraction had only been discovered by Laue in May of 1912, but that it had been quickly followed by the work of the father and son team William Henry and William Lawrence Bragg (see Sect. 1.8) [63, p. 320]. Out of the Bragg studies came the invention of the critical Bragg spectrometer and Bragg's law relating the wavelength λ of the X-rays to the planar spacing d in the crystal being studied and the angle of the scattered radiation ϑ:

$$n\lambda = 2d \sin \vartheta, \tag{5.1}$$

where n is the order of the diffracted waves [152, p. 59].

In our discussion of the work of Barkla and Sadler, we noted their discovery of what they referred to as secondary X-radiation in their studies of radiation absorption. It was not yet possible to measure wavelengths, as Barkla noted in 1911. Their discovery of secondary radiation was based on the penetrability of the incoming radiation and the absorption coefficient (defined as λ) in the standard equation for the X-ray intensity $I = I_0 \exp\left(-\lambda x\right)$. This equation failed when secondary X-rays were emitted from the sample being studied [9]. Based on penetration, however,

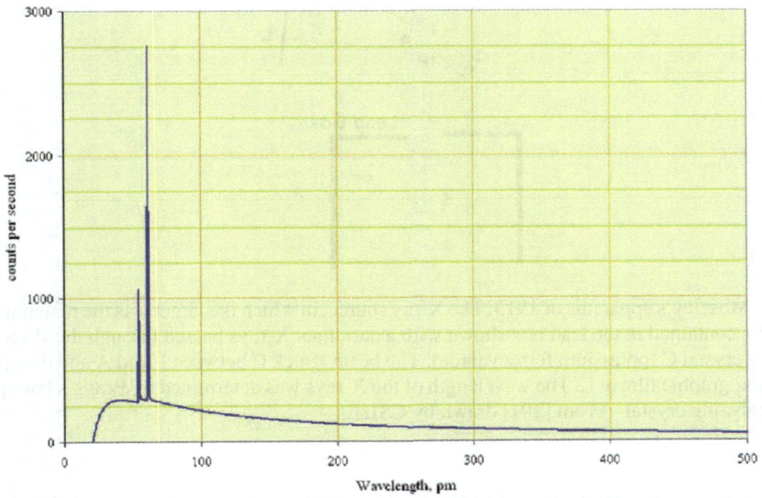

Fig. 5.1 Spectrum of X-rays emitted by a rhodium target at 60 kV. The continuous curve is due to bremsstrahlung. The spikes are characteristic K lines for rhodium. Public domain

Barkla realized that two types of radiation had been discovered, which he denoted by K and L, according to their penetrability [10].

In the winter of 1913, Moseley and Darwin had their X-ray equipment in place and by May they were using a crystal-based spectroscope [238, p. 82]. They found a continuous X-ray spectrum over a wide range of frequencies. But the Braggs had identified characteristic peaks in the spectrum, and a little later, Moseley and Darwin also found the peaks [63, p. 320].

Figure 5.1 shows a plot of the X-ray spectrum from rhodium at 60 kV. That is, the electron beam energy inducing the X-rays was 60 kV. The peaks are the K_α (higher) and K_β (lower) radiation for rhodium. The L series peaks are at a longer wavelength and are not shown in this graph [201].

When the electrons in the beam inducing the X-rays strike the target, most of the electron interactions with the atoms of the target release bremsstrahlung (braking radiation), which results when the electron trajectories are bent as they pass the nuclei of the target atoms. Acceleration of a charged particle causes emission of electromagnetic radiation.

The peaks in the X-ray spectrum, which are characteristic of the target material, are reminiscent of the characteristic spectral lines of an atom. It was known from Bohr's model of the hydrogen atom that this spectrum resulted from internal electron transitions from one orbit to another. Were the X-ray peaks also the result of electron transitions? As Moseley pointed out, Thomson had suggested that there was good reason to believe that the X-ray spectra came from the innermost ring of electrons [201, 270].

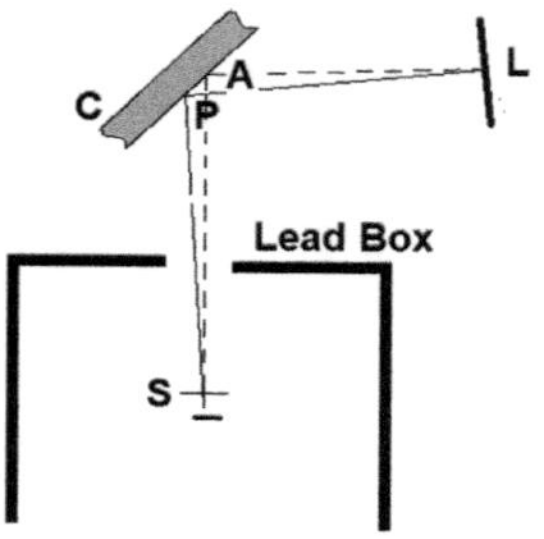

Fig. 5.2 Moseley's apparatus of 1913. The X-ray source, in which the target was the material under study, was contained in the lead box shown with a dark line. X-rays passed through the slit S to the analyzing crystal C (potassium ferrocyanide). The beam struck C between P and A and then passed to the photographic film at L. The wavelength of the X-rays was determined by Bragg's law applied to the analyzing crystal (From [201] drawn by CSH)

According to the Bohr model of the hydrogen atom, the frequencies of the characteristic spectral peaks should shift as the nuclear charge changes. If there was a shift in the X-ray peak frequencies with atomic number, the implication would be that the atomic number determines the nuclear charge. Moseley expected to get accuracies of at least one part in a thousand and hence to be able to detect a change in X-radiation wavelength with atomic number [238, p. 81].

Moseley carried out the experimental work himself with tremendous energy. Night-long sessions are not uncommon in experimental work, but Moseley seemed almost to have preferred them. [63, p. 320], [238, p. 82] Fig. 5.2 shows a diagram of the apparatus Moseley used in his experiments of 1913. The X-ray source, in which the target was the material under study, was contained in the lead box shown in outline. The X-rays passed through a slit S and then out of the lead box to the analyzing crystal C, which was potassium ferrocyanide. With equal distances S, (P, A), L the slightly divergent beam on C refocused on L. The angle ϑ in Bragg's law (5.1) was easily found from $2\vartheta = SAL$ or $2\vartheta = SPL$.

As an experimental result, Moseley plotted frequency ν against the square of the atomic number Z. (Our notation for the atomic number is the modern Z. Moseley used N.) The data plotted in Fig. 5.3 are fit by the equation[1]

$$\nu = \left(\frac{3}{4}\nu_0\right)(Z-1)^2\,, \tag{5.2}$$

in which ν_0 is Rydberg's wave number $N_0 = \nu/c = 109\,720$. This is a strictly experimental result.

[1] Any experimentalist will be duly impressed by the data presented in this plot. They are from Moseley's original paper.

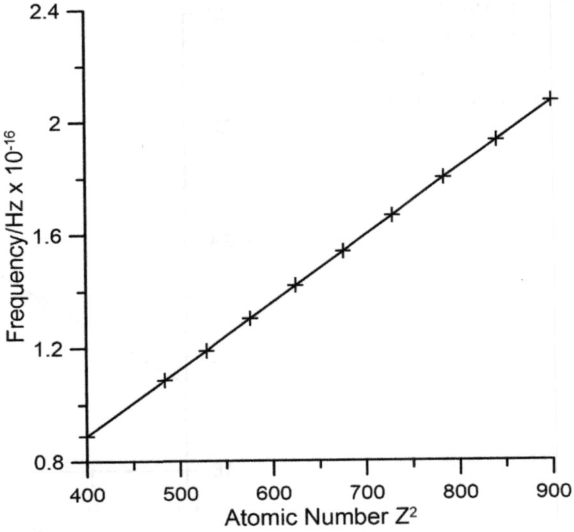

Fig. 5.3 Moseley's data of 1913 plotted as frequency versus atomic number squared. [201] (Plotted by CSH)

Moseley then defined the quantity

$$Q_{\mathrm{K}} = \sqrt{\frac{\nu}{\frac{3}{4}\nu_0}}.$$

The plot of Q_{K} as a function of the atomic number Z should then yield a straight line, as can be seen in Fig. 5.4. These data are fit by the equation

$$Q_{\mathrm{K}} = Z - 1. \tag{5.3}$$

The quantity Q_{K} then advances in value by unity as the atomic number is increased by a single digit.

According to Moseley: "We have here a proof that there is in the atom a fundamental quantity, which increases by regular steps as we pass from one element to the next." This quantity, he pointed out, could only be the charge on the central nucleus. The magnitude of the frequency of the characteristic X-ray lines increases regularly with this quantity.

To get a better understanding, he then turned to the physics. The angular momentum of an electron around the nucleus is

$$m\omega^2 r = \frac{e^2}{r^2}(z - \sigma_{\mathrm{n}}), \tag{5.4}$$

where σ_{n} represents the effect of the outside electron rings on the inner ring. Moseley provided estimates for these quantities, but decided to neglect them compared to the

Fig. 5.4 Moseley's data of 1913, plotted as the function $Q_K = \sqrt{\nu/\frac{3}{4}\nu_0}$ versus Z [201] (Plotted by CSH)

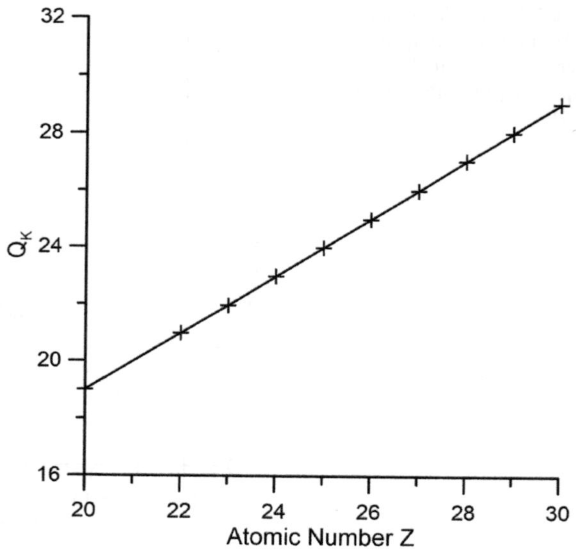

value of the atomic number Z. The σ_n values Moseley provided ranged from 0.25 to 2.81. From (5.4), it follows that $\omega^2 r^3$ is proportional to Z, so that

$$\left(\omega^2 r^3\right)_{Z+1} - \left(\omega^2 r^3\right)_Z \tag{5.5}$$

is a constant, as the atomic number Z changes by unity. Experimentally, Moseley had shown (in Fig. 5.4) that $\nu^{1/2}$ is a constant as the atomic number varies by unity, whence

$$\nu_{Z+1}^{1/2} - \nu_Z^{1/2} \tag{5.6}$$

is also a constant. Therefore,

$$\left(\frac{\omega^2 r^3}{\nu^{1/2}}\right)_{Z+1} - \left(\frac{\omega^2 r^3}{\nu^{1/2}}\right)_Z \tag{5.7}$$

is constant, as Z changes by unity.

The frequency ν here is the frequency of the characteristic radiation, while ω is the rotational frequency of the electron. How are they related? Bohr faced this same problem. In Sect. 4.2.2, we noted that Bohr simply chose $\nu = \omega/2$. In this context, then, Moseley is justified in taking $\nu \propto \omega$. Then $\omega^{3/2} r^3$ remains constant as Z changes by unity, and therefore $\left(\omega^{3/2} r^3\right)^{2/3} = \omega r^2$ is also a constant. Since the electron mass is constant, Moseley had then established that the angular momentum $m\omega r^2$ was a constant when we change the atomic number by unity. As he pointed out, this was experimental verification of what was first proposed by Nicholson and subsequently by Bohr (see Sect. 4.2.2).

Then Moseley turned to Bohr's result, which is, including σ_n,

$$\nu = \left(\frac{1}{1^2} - \frac{1}{2^2}\right) \frac{2\pi^2 e^4}{h^3} (Z - \sigma_n)^2 , \tag{5.8}$$

where e is the charge on the electron and h is Planck's constant. He compared this with the result from his experiments (5.2) and noted that they were identical if $\nu_0 = 2\pi^2 e^4/h^3$. The numerical values were, as Moseley pointed out, very close, while Bohr assumed them to be identical in his explanation of the Balmer series.

Moseley thus wrote: "This numerical agreement between the experimental values and those calculated from a theory designed to explain the ordinary hydrogen spectrum is remarkable, as the wavelengths dealt with differ by a factor of about 2000." Moseley admitted that he had no explanation for the faint K_β line, nor for the L lines [201].

In his second paper, which appeared in 1914, Moseley noted that more than 30 elements had been investigated, making it possible to predict with confidence the principal lines from aluminum to gold. The general experimental method was the same as the one used in the first (1913) part of this research. However, the apparatus was improved. A diagram of the second apparatus is shown in Fig. 5.5.

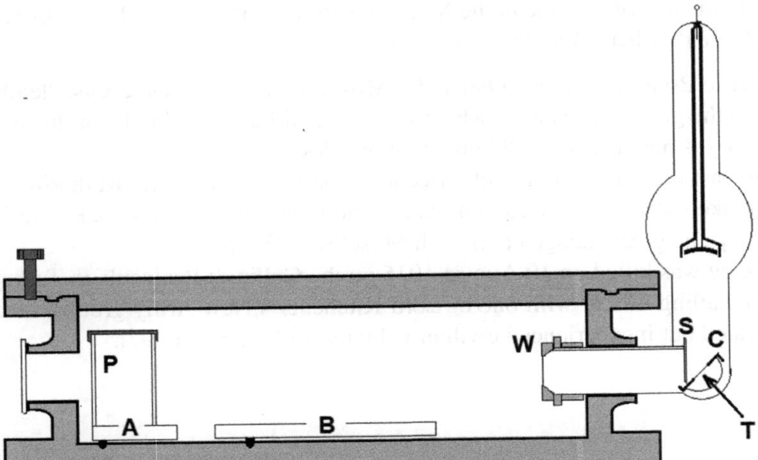

Fig. 5.5 Moseley's apparatus of 1914. The glass tube contained the electron beam and the target under study. Arms perpendicular to the picture were to transfer targets. The plate A held the photographic film P and rested on three steel balls, as did plate B with the potassium ferrocyanide analyzing crystal. S is the defining slit for the X-rays and W is the window transparent to X-rays. The apparatus was enclosed in an iron box 30 cm in diameter and 8 cm high. The lid was airtight (From [202] Drawn by CSH)

The new apparatus was specially designed and constructed for these experiments. The glass tube containing the X-ray source (the target under study) had two arms perpendicular to the picture to accommodate the carriage C and transfer mechanism for the target, which involved "a silk fishing-line wound on brass bobbins." The plate A held the photographic film P and rested on three steel balls. The plate B, mounted as A, held the analyzing crystal. S was the defining slit for the X-rays and W was the window covered with goldbeater's skin, which is transparent to X-rays.[2] Except for the X-ray tube and target carriage, the apparatus was enclosed in an iron box 30 cm in diameter and 8 cm high. The lid was airtight [202].

From his experiments Moseley then concluded: [202]

1. Every element from aluminum to gold is characterized by an integer Z, which determines its X-ray spectrum.
2. This integer Z, the atomic number of the element, is identified with the number of positive units of electricity contained in the atomic nucleus.
3. The atomic numbers of all elements from Al to Au have been tabulated on the assumption that Z for Al is 13.
4. The order of the atomic number is the same as that of the atomic weight, except where the latter disagrees with the order of the chemical properties.
5. Known elements correspond to all the numbers between 13 and 79 except three. There are here three possible elements undiscovered.
6. The frequency of any line in the X-ray spectrum is approximately proportional to $A\,(Z - b)^2$, where A and b are constants.

In a letter to Bohr on 16 November 1914, Moseley wrote that his results, "lend great weight to the general principles which you use, and I am delighted that this is so, as your theory is having a splendid effect on physics."

In his last interview Bohr said: "Because you see, actually the Rutherford work was not taken seriously. We cannot understand today, but it was not taken seriously at all. [...] The great change came with Moseley" [238, p. 85].

Moseley was killed on 10 August 1915 on the Gallipoli Peninsula in Turkey. He was a signalling officer with one of Lord Kitchener's New Army groups, made up of dedicated but inexperienced civilian volunteers [238, pp. 96–97].

[2] Goldbeater's skin is the processed outer membrane of the intestine of an ox, valued for its strength against tearing.

5.2.3 *Franck and Hertz*

In 1914 James Franck[3] (1882–1964) and Gustav Hertz (1887–1975) conducted an experiment which stands historically as a crucial experimental verification of the Bohr model.

In 1911, at the University of Berlin, Franck and Hertz began investigating the ionization potentials of atoms and molecules by studying inelastic collisions of the atom or molecule with electrons. Their idea was to vary the energy of a beam of electrons passing through a low density gas of the element in question and to measure the current from the electrons. When the ionization energy was reached they expected the current to drop because energy had been lost by electrons as they ionized the atoms or molecules of the gas. Franck and Hertz had publishable results from a set of experiments on mercury in 1914 [114, 115].

Figure 5.6 is a schematic of the first apparatus used by Frank and Hertz. This consisted of a central platinum wire, concentric to platinum mesh cylinder and a platinum solid cylinder. The mesh cylinder, with a radius of 4 cm, was inside the solid cylinder and separated from the solid cylinder by 1 to 2 mm. In the experiment, a current was passed though the central wire causing it to glow and release electrons. A potential between the central wire and the mesh cylinder provided the energy input to the electrons. The solid cylinder was connected to earth through a galvanometer to measure the current from the electrons passing through the mesh and striking the cylinder. This compound cylinder was mounted in a closed quartz container, which was connected to a pump by a U-tube containing mercury at the lowest point. In this way the pressure of the mercury vapor in the quartz container could be controlled and measured.

All electrical connections were fused to the quartz to prevent gas leakage. The apparatus was mounted in a controlled paraffin heat bath and the temperature was between 110 and 115 °C. Franck and Hertz measured a series of peaks in the galvanometer current, spaced 4.9 V apart in the wire to grid potential [114].

During the experiment, Franck and Hertz also measured a spectral emission at a wavelength $\lambda = 2536$ Å. Using this wavelength in Planck's formula $\mathcal{E} = h\nu = h\,(c/\lambda)$, they obtained an energy of 4.84 eV, which they interpreted as the energy emitted by a free electron falling back into an orbit in the mercury atom.[4] This was close enough to the 4.9 V they had measured in the galvanometer current to justify taking it as confirmation of their measurement of the ionization potential [114].

---

[3] Franck exhibited moral courage as one of the first to take a stand against the racial laws in Nazi Germany, resigning from the University of Göttingen in 1933. He then accepted a position at the Johns Hopkins University in Baltimore, and in 1938 moved to the University of Chicago, where he later became part of the Manhattan Project [238, pp. 191–192]. In 1945 (two months before Hiroshima), he and a group of scientists drafted what is known as the "Franck Report" to the War Department, urging an open demonstration of the atomic bomb in some uninhabited locality as an alternative to the military decision to use the weapon without warning in the war with Japan. Although it failed in its objective, this document remains a fundamental example of the resistance of scientists to use science in works of destruction [142, p. 423].

[4] Our calculation with modern values for c and h yields $\mathcal{E} = 4.889$ eV.

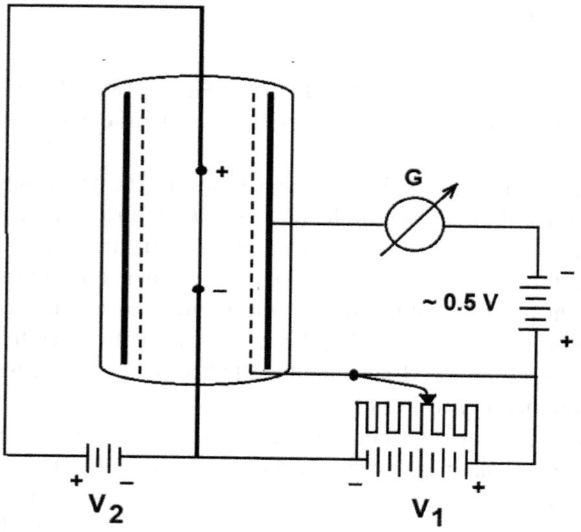

Fig. 5.6 Franck and Hertz's first apparatus. The gas studied was mercury vapor. The potential V_1 is the electron accelerating potential. The potential V_2 heats the central wire inside the quartz tube, thereby providing the electrons. *G* is the galvanometer to measure the electron current from the central wire to the outside cylinder (From a meeting at the PTR, 24 April 1914. Drawn by CSH)

Franck and Hertz subsequently modified their apparatus and conducted a similar set of experiments, also in 1914 [115]. The left panel of Fig. 5.7 shows the apparatus Franck and Hertz used in this second set of experiments [115].

The right panel of Fig. 5.7 shows a graph of their data. The apparatus had changed considerably, with the galvanometer connected directly to the mesh. A flame heater kept the temperature at about 150 °C. The data, which were unchanged from those obtained in the previous experiments, showed peaks spaced at 4.9 V. The sketch of the apparatus appeared in the English translation of the original paper by Franck and Hertz [264, pp. 160–166].

The dynamics of the electrons colliding with the low density mercury vapor was not simple. In general, a certain statistical fraction of the electrons undergo collisions with the mercury atoms, and even when the electrons have an average energy sufficient for an inelastic collision, not all collisions will be inelastic. The energy of the electrons leaving the glowing wire was another unknown, and the wire potential varied over the wire length [114]. Nevertheless, Franck and Hertz could understand the peaks in the graph in the right panel of Fig. 5.7 as the result of a specific energy barrier of 4.9 eV which they interpreted as the ionization potential of mercury.

Apparently, Franck and Hertz were unaware of Bohr's 1913 publication of his atomic model, and in both of the 1914 publications they interpreted their data as resulting from ionization of the mercury. Bohr, however, interpreted these data differently. In 1915 he pointed out that the ionization energy for mercury could be obtained from Paschen's single spectral line series for mercury, which contains lines

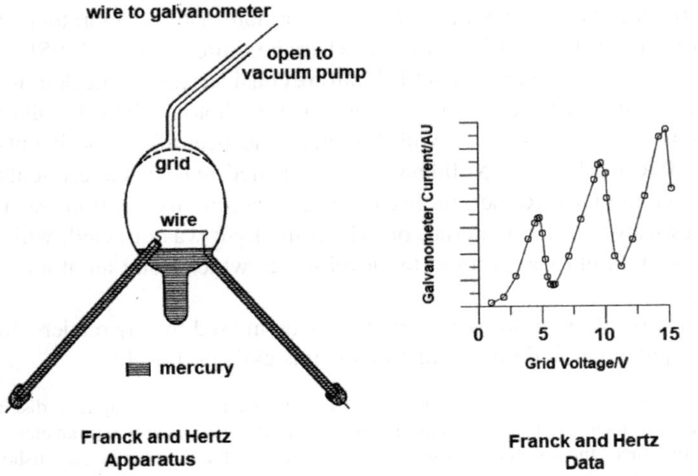

Fig. 5.7 Franck and Hertz's second apparatus. In the left panel is a drawing of the apparatus used by Franck and Hertz in the second experiments they reported in 1914 [115]. In the right panel is a plot of their data. (Adapted from [264], drawn by CSH)

at 1850, 1439, 1269 Å. Using Bohr's theory this yields a limit of 10.5 eV for ionization. Bohr believed that the line that Franck and Hertz discovered, of wavelength 2536 Å, must correspond to an internal transition of the electron and not an ionization.

Franck and Hertz (finally) accepted Bohr's argument in 1919 and wrote a review [116, 188, pp. 367–369]. The beautiful experiment of Franck and Hertz was thus a marvelous demonstration of the existence of Bohr-like orbits in mercury.

Note that both Franck and Hertz were in the German Army in WW I after 1914. This limited their scientific activities until after 1918.

5.3 The Einstein Photon

5.3.1 Experiments by Millikan

Robert A. Millikan (1868–1953), the son of Rev. Silas Millikan and Mary Jane Andrews, was born in rural Morrison, Illinois. He graduated from high school in Iowa and in 1886 entered Oberlin College, where he studied Greek and mathematics. After graduating in 1891 he remained at Oberlin for two years to teach elementary physics. Millikan received a fellowship in physics at Columbia and obtained his Ph.D. there in 1895. On the advice of his professors, he spent a year 1895–1896 in Germany at the universities in Berlin and Göttingen. Then, on the invitation of Albert A. Michelson (1852–1931), Millikan took up a position at the University of

Chicago. In 1921 he became director of the Norman Bridge Laboratory of Physics at the California Institute of Technology, where he retired in 1945 [195].

Traditionally the experiments of Millikan are considered to be the definitive experiments that established the validity of the Einstein photon. The difficulty with this claim lies in the fact that, even with his clear and beautiful experiments leading up to the 1916 publication, Millikan actually denied that his experiments verified Einstein's photon idea. Rather, in this 1916 publication, his position was that there must be resonators in the material, on which the light was directed, which gained energy from the light waves up to the level $h\nu$, at which point an atom ejected an electron[5] [193].

Millikan's position in this paper can hardly be missed by any reader. On the first page of the paper we find this evaluation of Einstein's proposal:

> This hypothesis may well be called reckless first because an electromagnetic disturbance which remains localized in space seems a violation of the very conception of an electromagnetic disturbance, and second because it flies in the face of the thoroughly established facts of interference [193].

Millikan wrote, in the last section of his 1916 paper, and repeated in his book published in 1917, that "Einstein himself, I believe, no longer holds to it" (the idea of the photon) [193, 194, 261]. He reversed this position, however, in his 1950 autobiography. There he said that his experiments reported in 1916 permitted no other interpretation than that which Einstein had ventured in 1905 [196, 261]. Millikan's confession in 1950 is the one most often quoted.

The paper we discuss here is entitled *A Direct Photoelectric Determination of Planck's "h"*. This was a culmination of the work he began in 1905, which may indicate that he had difficulties with Einstein's ideas as soon as he encountered them. Millikan was at least very familiar with the predictions in Einstein's paper and noted that none of them had been thoroughly tested, except for the maximum velocity of the photoelectrons from a specific excitation frequency. He pointed out, however, that the validity of this had been denied by Carl Ramsauer (1879–1955) in 1914 [233].

Millikan specifically indicated that the linear relationship between the stopping potential (Π in Einstein's notation, see Sect. 3.3.8) and the exciting frequency ν, viz.,

$$\Pi = \left(\frac{h}{q_{\mathrm{e}}}\right)\nu - \frac{P}{q_{\mathrm{e}}}, \tag{5.9}$$

where P is the energy required to remove the electron from the surface of the material being studied (the work function), had not been established. Millikan's oil drop experiment of 1913 had yielded $q_{\mathrm{e}} = 4.774 \times 10^{-10}$ statcoulombs (compared to the present accepted value of 4.803×10^{-10} statcoulombs) [192]. If he could establish the linear relationship in (5.9), he would then be able to calculate the constant of action h.

[5] Millikan's position here, we may note, was strictly Thomsonian regarding the role of electrons.

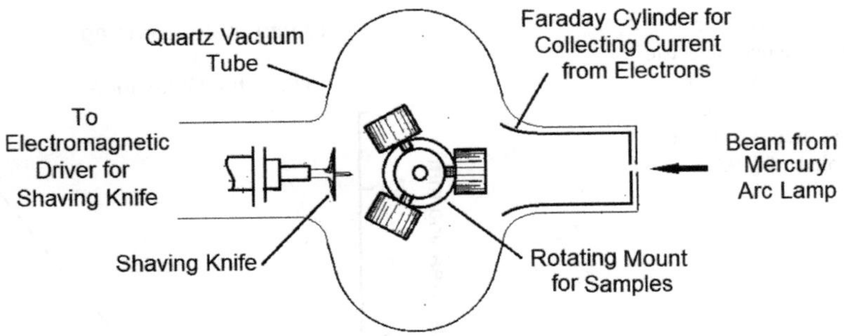

Fig. 5.8 Millikan's apparatus for determining Planck's constant. A detailed drawing of the apparatus appears in [193]. Here we include only the principal parts of the apparatus (Drawn by CSH)

Just in terms of finding the exact value of h, this was then a critical experiment. Holton notes that this experiment richly deserved to be cited as part of Millikan's Nobel Prize award in 1923, attributed "for his work on the elementary charge of electricity and the photoelectric effect."

Figure 5.8 shows the main features of Millikan's apparatus. All the working parts necessary to carry out the experiment, except the mercury arc lamp that supplied the light beam and the instruments for measuring voltages and currents, were contained within a quartz vacuum vessel. This in itself is remarkable. However, it was a serious consequence of Millikan's conviction that any film adhering to the surface of a material will adversely affect the experimental results.

At the center of the apparatus was rotating mount with a set of three cylindrical samples. The sample on which the measurements were to be made was first rotated to be coaxial with the shaving knife. The revolving knife was then moved to the face of the sample by an external electromagnet and the face shaved to remove any possible film. The knife was then backed away and the cleaned sample lined up with the port through which the light beam would enter from a mercury lamp. The experiment was conducted in some haste to obtain data before a new film could form.

A Faraday cylinder collected the electrons emitted (immediately) from the surface when the beam was turned on. The current was measured as a function of the potential difference applied between the sample and the Faraday cylinder. The current measurements had to be carried out with the greatest precision available in order to accurately determine the voltage at which the current had been stopped. For this, Millikan used a Dolezalek electrometer invented by the Hungarian/German physicist, Friedrich Dolezalek (1873–1920). A detailed description of this instrument may be found in [113]. The electrometer measures charge accumulated over a selected time interval from the current source to be measured. The resulting deflection of the needle or vane of the instrument is measured by a light beam incident on a tiny suspended mirror and is given in distance (mm) traversed by the reflected light beam on a scale. Millikan selected an accumulation time of 30 s for his measurement of the very small photoelectric current near the stopping potential.

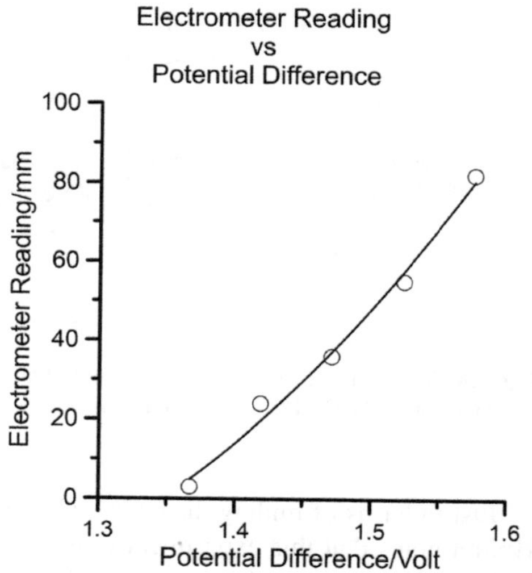

Fig. 5.9 Electrometer reading of the photoelectric current from wavelength $\lambda = 4,047$ Å near the stopping potential (Data from Table 1 of [193]

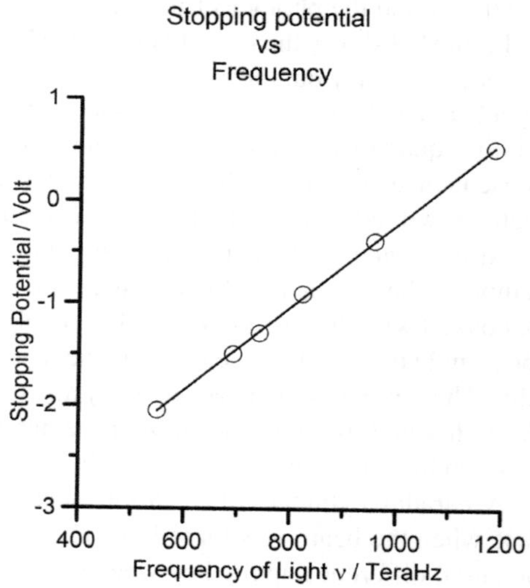

Fig. 5.10 Results of Millikan's experiment. This is a plot of experimental data for Eq. (5.9). The slope of this line is h/q_e. Data read by eye from Fig. 5 [193] (Plotted by CSH)

Figure 5.9 shows the plot to determine the stopping potential for wavelength $\lambda = 4,047$ Å. Data are from Table 1 of Millikan's paper [193].

Millikan's final analysis was based on a plot of (5.9) from his experimental results for the stopping potential at each wavelength. Figure 5.10 plots Millikan's experimental data for the stopping potential as a function of the light frequency. This linear plot clearly matches the prediction of equation (5.9).

This work may be considered as an example of experimental genius. Possible difficulties, like films on the samples, were encountered and discussed at length. The final section of the paper summed up as follows:

1. Einstein's photoelectric equation has been subjected to very searching tests and it appears in every case to predict exactly the observed results.
2. Planck's h has been photoelectrically determined with a precision of about 0.5 per cent and is found to have the value

$$h = 6.57 \times 10^{-27} \text{ erg s.}$$

The penultimate section of the paper was entitled *Theories of Photo Emission* and took up 5 of the 33 pages in the paper. Millikan began this section as follows:

> Perhaps it is still too early to assert with absolute confidence the general and exact validity of the Einstein equation. Nevertheless, it must be admitted that the present experiments constitute very much better justification for such an assertion than has heretofore been found, and if that equation be of general validity, then it must certainly be regarded as one of the most fundamental and far reaching of the equations of physics; for it must govern the transformation of all short-wave-length electromagnetic energy into heat energy. Yet the semi-corpuscular theory by which Einstein arrived at his equation seems at present to be wholly untenable.

Millikan concludes this penultimate section by saying:

> This is little more than Planck's theory with the possibility of a corpuscle being emitted from an atom with an energy greater than $h\nu$ eliminated for the sake of reconciling it with the experimental facts above presented.

In this section, Millikan considered the ideas of J.J. Thomson and of Planck, as well as his own oil drop experiments, in which he certainly observed discontinuities of charge, but not of energy.

5.3.2 Compton's Experiments

The experimental work by Arthur H. Compton (1892–1962) was actually the most important in resolving the issue of the photon.

Compton graduated from the College of Wooster in 1913, where his father, a Presbyterian minister, was professor of philosophy. He then received his Ph.D. from Princeton University in 1916. After a year as an instructor at the University of Minnesota and two years with Westinghouse Electric and Manufacturing in Pittsburgh as a research engineer, he became a Fellow of the National Research Council and went to work with Rutherford in Cambridge in 1919. Rutherford had just returned to Cambridge, succeeding J.J. Thomson as the Cavendish Professor and Director of the Cavendish Laboratory. There Compton began studies of the scatter of γ-rays, observing an increase in the wavelength of the scattered γ-rays compared to the initial beam. He attributed this increase in wavelength to the Doppler effect.

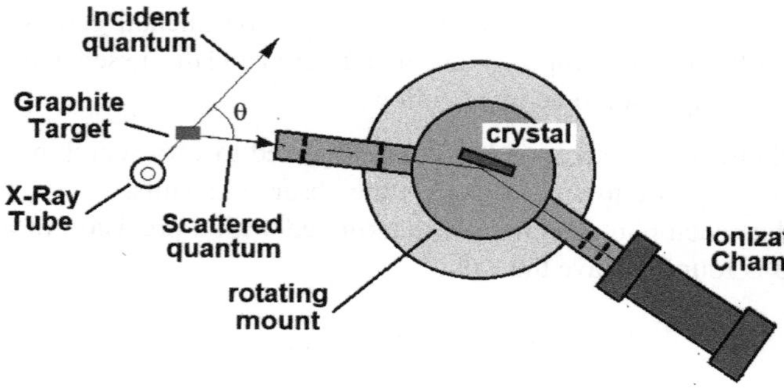

Fig. 5.11 Compton's apparatus. Two parts are pictured: (1) The X-ray tube including the graphite target and (2) the Bragg spectrometer. The Bragg spectrometer was inside a lead box. The first section of the collimator was outside the lead box. The crystal in the spectrometer was fixed on a rotating mount so that the crystal and ionization chamber could be rotated independently. The wavelength of the scattered beam was determined by Bragg's law from the angular spectrometer measurement (From Fig. 1 [57], Drawn by CSH)

After a year at the Cavendish Laboratory, Compton accepted the Wayman Crow Professorship at Washington University in St. Louis, Missouri, where he intended to continue his research into radiation scattered from electrons in matter. Because he needed to make precise measurements of the spectra of both the primary and the scattered radiation, he brought a Bragg spectrometer with him from England. Using a steady X-ray beam, the Bragg spectrometer allowed him to make precise measurements of the wavelengths of both the primary and scattered beams. He was then prepared for a detailed study of radiation scattered by electrons.

Figure 5.11 is a schematic drawing of Compton's basic apparatus. Compton obtained his first X-ray data in December of 1921. He used molybdenum K_α X-rays with wavelength $\lambda = 0.708$ Å as the primary beam. At a scattering angle of 90°, he detected a peak with wavelength very close to that of the primary beam and a less intense peak with wavelength $\lambda' = 0.95$ Å. He interpreted this less intense peak as the scattered beam. For the ratio of the wavelength of the scattered beam to that of the primary, he obtained $\lambda/\lambda' = 0.75$.

From an analysis based on energy conservation and the Doppler effect, he obtained a theoretical prediction of $\lambda/\lambda' = 0.74$. Of course, he published this, in spite of the fact that the shift in wavelength was so high. By October of 1922, however, he realized his error. The major peak at 90° was the peak produced by the scattered X-rays, and the ratio of the wavelengths was $\lambda/\lambda' = 0.969$. He used the Doppler effect once again, but this time he appealed to conservation of momentum in his analysis. The result of this theoretical calculation was $\lambda/\lambda' = 0.966$. The agreement between theory and experiment was marvelous once again.

He discussed the result with a colleague in his department, Professor George E. M. Jauncey (1888–1947), and realized that the theoretical analysis should include both energy and momentum conservation [56]. The analysis then produced the familiar result in modern physics texts, which is

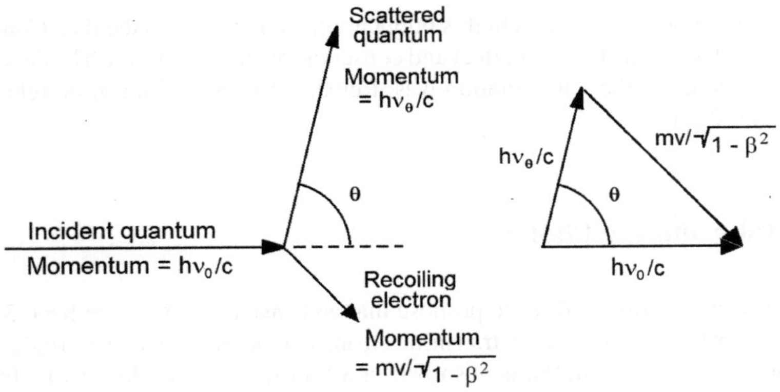

Fig. 5.12 Compton effect. The left panel shows a diagram of the collision and the right shows the momentum triangle (From Fig. 1 [56], drawn by CSH)

$$\Delta\lambda = \lambda' - \lambda = \left(\frac{h}{mc}\right)(1 - \cos\theta). \tag{5.10}$$

He submitted this result to *The Physical Review* on 13 December 1922 [56].

This is what is now known as the *Compton effect*. With momentum and energy conservation, this is the result of an elastic collision. The title of the publication was *A Quantum Theory of the Scattering of X-Rays by Light Elements*. The momentum (vector) triangle as it appeared in Compton's paper leaves no room for doubt regarding Compton's understanding of the physics of the collision. We have redrawn this in Fig. 5.12.

The radiation quantum with momentum $h\nu_0/c$ undergoes an elastic collision with an electron at rest.[6] The radiation quantum leaves the collision point at an angle θ with the horizontal and a momentum $h\nu_\theta/c$, while the electron leaves the collision point with a (relativistic) momentum $mv/\left(1-\beta^2\right)^{1/2}$, where $\beta = v/c$.

There was, however, a sticking point. Compton only measured the X-rays emitted at the angle θ. Figure 5.12 assumes that a single photon with momentum $h\nu_\vartheta/c$ leaves the collision point simultaneously with the electron. But could it be that the electron simply absorbs the photon and emits the radiation in accordance with classical bremsstrahlung? Compton carried out a detailed study of this situation. It could not be entirely eliminated. The study was of scatter from light elements and Compton pointed out that more tightly bound electrons in heavy elements might yield different results.

[6] The electron is actually bound to an atom. But the energy of the X-ray photon was considerably higher than the electron binding energy. This is the reason Compton chose low atomic weight atoms as targets.

We should note that the data in this publication are identical to the data Compton had analyzed using the Doppler effect and conservation of momentum. The data were not in question, but the interpretation was. Remarkably, he still made no reference to Einstein [261].

5.4 Bohr and the Photon

It would seem natural for Bohr to propose that an Einstein photon (see Sect. 3.3) is emitted when the hydrogen atom transitions from a higher to a lower electronic state, and that, conversely, a photon is absorbed in a transition from a lower to a higher state. Bohr, however, did not introduce the photon to explain transitions between his proposed stable orbits in the atom simply because he did not accept Einstein's ideas about the photon. Bohr's position on the Einstein photon is rather clearly revealed in his Nobel lecture of 1922, when he said:

> In spite of its heuristic value, [...] the hypothesis of light-quanta, which is quite irreconcilable with so-called interference phenomena, is not able to throw light on the nature of radiation.

Although in 1922 he was convinced that Einstein's proposal was in error, he was by then drawing close to the end of his objections regarding the Einstein photon. The physics involved in the history of the conflict surrounding the concept of the light quantum is rather interesting, so let us pursue it a little further here.

5.4.1 Ideas of Bohr, Kramers, and Slater

John C. Slater (1900–1976) received his Ph.D. from Harvard in 1923 and a Harvard Sheldon Fellowship for study in Europe. He went first to Cambridge and then to Copenhagen, where he arrived in November of 1923.

He had formulated a theory in which light quanta were guided by electromagnetic waves. This was something of considerable interest to Bohr and Hendrik A. Kramers (1894–1952). However, Slater was never completely successful in his attempts to convince Bohr and Kramers of his theory and chose to be flexible for the publication that appeared in 1924 [16, 17]. That paper, which is often known simply by the initials of the three authors Bohr, Kramers, and Slater (BKS), used the basic concepts of Slater's theory with the quanta dropped. The BKS paper is now only of historical interest.

In the paper the authors outlined a formulation of the mechanism of interactions between and among atoms which recognized the quantum theory of 1923 but was devoid of any of the Einstein light quanta. They linked the classical and quantum theories via Bohr's *correspondence principle*,[7] which they identified as a general

---

[7] The *correspondence principle* is the claim that the results of quantum mechanics must be the same as those from classical mechanics in the limit at which we expect classical mechanics to apply. It

conjugation of the transitions between stationary states of an atom with one of the harmonic oscillator components into which the electrical moment of the atom could be resolved. They also linked the interaction between free electrons and radiation by relating the change in wavelength of the scattered rays and the classical Doppler effect, roughly as Compton had done.[8]

Then they introduced their fundamental assumption that an atom in a stationary state communicates with all other atoms through a spatiotemporal mechanism originating from virtual harmonic oscillators corresponding to the various possible transitions to other stationary states. A transition in a particular atom depends on the initial state of the atom and on the states of atoms with which it is in communication through this virtual radiation field. This was to merge into the picture produced by the classical radiation theory. Since these processes could only be described probabilistically, conservation of energy and momentum became statistical properties of the interactions and were not to be applied to individual interactions.

The remainder of the paper was devoted to examples and culminated with an illustration of the transition to Maxwell's electromagnetic theory. Some inspiration for the proposed transition theory was derived from Einstein's ideas about transitions in collections of molecules exposed to electromagnetic radiation [101, 215, pp. 405–407].

Shortly after the publication of BKS, two experimental papers were published that ruled out the ideas ventured in BKS. The first of these was the experiment conducted by Walther Bothe (1891–1945) and Hans Geiger, while the second was conducted by Compton[9] and his student Alfred W. Simon.

5.4.2 *Bothe and Geiger's Experiment*

Figure 5.13 shows the central part of the apparatus built by Bothe and Geiger. In the experiment two opposing needle probes were arranged on either side of a target for the Compton effect. An X-ray beam passed between the probes. In Fig. 5.13 this is the line from top to bottom through the center. In the upper left of Fig. 5.13 is a drawing of the face of the photon counter ($h\nu$ counter). The aluminum (Al) on the face of the counter is the target.

The drawing on the right of Fig. 5.13 shows the Compton effect inside the Al target. The electron entering the electron counter will correspond to a scattered photon entering the $h\nu$ counter. When this scattered photon strikes a gas atom in the $h\nu$ counter, another Compton effect will occur and the electron will be detected by the probe.

---

was Niels Bohr who originated the correspondence principle and apparently often considered the limit to be that of large quantum numbers.

[8] We have already noted that this is not exactly what Compton had done.

[9] In 1923 Compton moved to the University of Chicago.

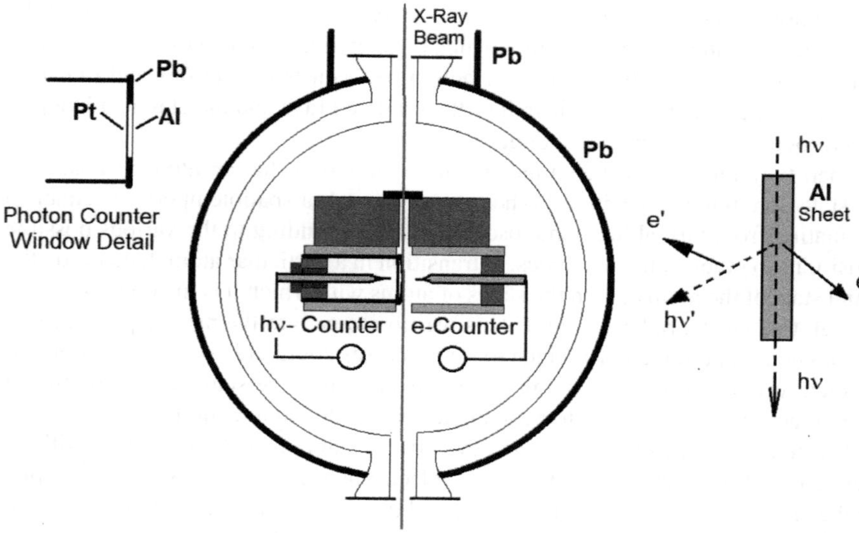

Fig. 5.13 The Bothe–Geiger experiment. An X-ray beam passes through the center of the apparatus striking the Al target on the window of the photon counter ($h\nu$ counter) resulting in a Compton effect. A detail of this window is shown in the left panel. The right panel is a diagram of the Compton effect in the Al. The photon from this Compton effect will induce a second Compon effect in the gas in the $h\nu$ counter, producing the electron that is counted by the $h\nu$ counter ([30] Drawn by CSH)

Each probe was connected to an electrometer. The connection to the electrometers is indicated by the open circles below the counters in Fig. 5.13. The output of each electrometer was recorded on one of two silver bromide films, which were moving past the electrometers at high speed. If the pulses recorded on the films coincided, this meant that the electron and the light quantum had been produced simultaneously. Simultaneity would establish conservation of energy and momentum at an atomic level, in accordance with the momentum triangle in Fig. 5.12. If the responses of the two electrometers were not simultaneous, the BKS theory would at least be credible.

The experimental work was laborious and time-consuming. For example, the length of film to be developed and analyzed was more than 3 km. And the analysis required almost a year. Bothe and Geiger were able to report, however, that the responses of the two electrometers were simultaneous to within 10^{-4} s. This alone was sufficient to refute the BKS proposal [29].

5.4.3 Experiment of Compton and Simon

In their experiment, Compton and Simon elected to study Compton scattering in a cloud chamber [59]. They were already aware from the data of Charles T.R.

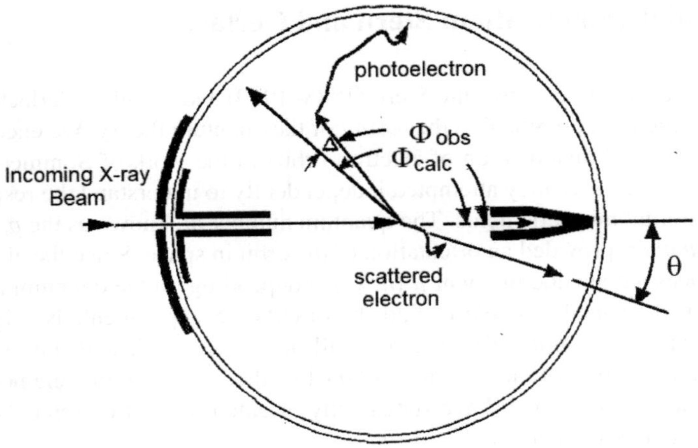

Fig. 5.14 Compton and Simon's apparatus. Only the cloud chamber, which was contained in a lead box, is shown in this drawing. The X-ray source, also contained in a lead box, is to the left of the drawing. The photon emerging from the collision with a longer wavelength than the X-ray is located with the angle Φ and the electron with the angle θ (From Fig. 4 in [57], drawn by CSH)

Wilson[10] (1869–1959) in Cambridge that a cloud chamber exposed to X-rays produced photoelectrons resulting from Compton collisions of X-rays with molecules in the air. These photoelectrons left chaotic tracks, which Wilson had called "fish tracks". The appearance of these tracks could be used to mark light quanta from the Compton collision, and the electron produced in the collision provided by the X-ray beam was collimated into a very thin needle.

Figure 5.14 shows only the cloud chamber portion of Fig. 4 from a presentation by Compton to a joint session of the AIP and the AAPT in 1961 [57]. Both the cloud chamber and the X-ray source were contained in lead boxes. In Fig. 5.14, the angle Φ locates the scattered photon and the angle θ locates the scattered electron. This is the notation used by Compton in his 1961 paper [57]. We have drawn the scattered electron and the photoelectron produced by the scattered photon as "fish" tracks, as they were drawn in Fig. 4 of Compton's paper. The beginning of the photoelectron track places the scattered photon experimentally at the angle Φ and the tangent line to the emerging scattered electron track places the line at angle θ. The small angle Δ indicates the error between the calculated angle of scatter of the photon and the observed angle.

With the experiments by Bothe and Geiger and by Compton and Simon there was no option for Bohr but to accept the reality of the photon. After the Bothe and Geiger, publication he finally accepted in a letter to Ralph H. Fowler (1889–1944) in Cambridge on 21 April 1925 [261] that his ideas expressed in BKS must be given as honourable a funeral as possible.

[10] Wilson was the inventor of the cloud chamber, for which he received the 1927 Nobel Prize in physics.

5.5 Spatial Quantization: Stern and Gerlach

The experiment conducted by Otto Stern (1888–1969) and Walther Gerlach (1889–1979) was crucial in establishing the reality of the quantum theory. We encountered the issue of spatial quantization of electron orbits in the work of Sommerfeld and Debye in Sect. 4.3.2, as they attempted independently to understand the response of atoms to external magnetic fields. The quantum number m, known as the *projection quantum number*, provided an orientation of the orbit in space. Since the electron in orbit produces a magnetic moment $\boldsymbol{\mu}\,(n, k, m)$ depending on the quantum numbers of the orbit, it seemed possible to find the orientation experimentally. Figure 4.3 illustrates this very simply. The electron will have one or other of the two orbits shown. The prediction of the Sommerfeld–Debye theory is then that elements with single valence electrons will have statistically oriented magnetic moments. Could this be detected experimentally?

Note that we will ignore the electron spin in our discussion. Our reason for doing this is based on the historical point at which the experiment was finally successful, which was 1922, while the electron spin was only proposed in 1925. We discuss this in Sect. 9.2.

Stern was the driving force with the initial idea, but Gerlach was a more accomplished experimentalist. Stern was born in Sohrau in what was then the Kingdom of Prussia. His family moved to Breslau in 1892 and he received his doctorate in physical chemistry from the University of Breslau in 1912. His dissertation was on the kinetic theory of osmotic pressure (Fig. 5.15).

His parents were sufficiently affluent to provide funds for postdoctoral study and Stern decided he would like to study with Einstein, who was then at the University of Prague. Einstein was willing to take Stern on as his first student. When Einstein was called back to the University of Zurich, Stern followed him and became a *Privatdozent* in physical chemistry at the Eidgenössische Technische Hochschule (ETH) in 1913.

Fig. 5.15 University of Breslau, 1900 (Unknown author. Image available from the United States Library of Congress's Prints and Photographs division. Public domain)

Fig. 5.16 Hans Poelzig Building (IG-Farben building) of Frankfurt University (Picture taken in June 2003 by Johannes Marx)

Fig. 5.17 University of Tübingen Alte Aula (*Photo*: Berthold Werner)

When Einstein was called to Berlin in 1914, Stern accepted a post as *Privatdozent* at the Johann Wolfgang Goethe University of Frankfort am Main (Fig. 5.16).

After military service, Stern worked with Walther Nernst (1864–1941) in Berlin and then returned to Frankfort where he became assistant to Born at the Institute for Theoretical Physics. He accepted a call as full professor at the University of Rostock in 1921 [118, 185, p. 433].

Gerlach was born in Biebrich am Rhein, Germany. He did his doctorate in physics at the Eberhard Karls University of Tübingen in 1912 and continued there as an assistant to Paschen until 1915. He wrote his habilitation at Tübingen in 1916. After military service he held a position in the physical laboratory at Farbenfabriken Elberfeld until 1920, then returned to academia, becoming *Privatdozent* and then extraordinary professor at the University of Frankfort am Main. In 1925 he was appointed to the chair previously held by Paschen [185, p. 436] (Fig. 5.17).

At the Institute for Theoretical Physics in Frankfort, Stern encountered the work of Louis Dunoyer (1880–1963) from 1911, which involved studies of sodium atom beams. He was impressed with the simplicity of the method and the possibility of making measurements directly on neutral atoms with large scale equipment. Born strongly encouraged Stern in his interests and in 1919 Stern and his student Elisabeth Bormann were measuring mean free paths and velocity distributions in beams of silver atoms [185, p. 434].

Gerlach arrived at the Institute for Experimental Physics adjacent to Born's institute in 1920. Gerlach had been trained by Paschen at Tübingen and Born was delighted to have him next door [118, 185, pp. 433, 436] Gerlach had also encountered Dunoyer's work in 1912 and had pursued some experiments without success. When he came to Frankfort, his interest was in trying to demonstrate diamagnetism in atoms of bismuth by the deflection of a beam of bismuth atoms in a strong inhomogeneous magnetic field. The force on a magnetic moment $\boldsymbol{\mu}$ in a magnetic field of intensity $\mathbf{H}$ and varying in the z direction is the rate of change of the dipole energy in the direction z, which is $\boldsymbol{\mu} \cdot (\partial \mathbf{H}/\partial z)$. The only way to get the magnetic dipole to move spatially is to make the magnetic field inhomogeneous. This could be accomplished by carefully shaping the poles of two permanent magnets and placing them close together

Magnetism in solids had been studied by Pierre Weiss (1865–1940) in 1913, resulting in a theory of magnetic susceptibility in solids. Weiss' model accounted for the magnetization of the material as well as the individual magnetic moments of the atoms. In Weiss' model, however, the magnetic moment of magnetized iron turned out to be about one fifth of a Bohr magneton. Then in 1920, Pauli attempted to bring in spatial quantization by conducting a statistical average over the projection quantum numbers m, but also concluded that the magnetic moment was much less than a Bohr magneton. The physical problem was not solved, but Pauli had introduced the term spatial quantization, which made the community aware of the concept, and that community included Stern.

In a seminar one afternoon, Stern recalled, one of the topics had been birefringence, which we discussed in Sect. 4.3.2. According to Lorentz's theory of the Zeeman effect, birefringence should be present whenever a material is located in a magnetic field. However, although common in liquids and solids, birefringence had never been observed experimentally in gases [117]. Any experimental study of spatial quantization in individual atoms could not, therefore, be based on spectroscopic studies of birefringence [118].

Stern was gaining valuable experience with atomic beam studies. This would place him in an excellent position to conduct an experiment on individual atoms in an atomic beam, which could be treated physically as gas atoms. He then needed only to conceive of a method for measuring the space quantization of the beam atoms without appealing to birefringence. According to Stern, the morning after the seminar was quite cold and he had awakened early. With no desire to get out of a warm bed he lay there thinking about the seminar question and had an idea for a possible experiment [117].

The fact that he was studying beams of silver atoms, we may suppose, influenced Stern's thinking. The silver atom has a single valence electron making it a model for hydrogen. The single valence atom rotates in the field of a shielded nucleus with total charge equal to that of the proton in hydrogen. In the Sommerfeld–Debye model of quantum orbits, the principal quantum number in the lowest state is then $n = 1$ and there is a single azimuthal quantum number $k = 1$. There are thus two possible values for the projection quantum number, which is $m = \pm 1$. The situation is then as illustrated in Fig. 4.3 of Sect. 4.3.2. The single valence electron must be in one of the states $(n, k, m) = (1, 1, 1)$ or $(n, k, m) = (1, 1, -1)$. In the presence of an external magnetic field of intensity H, the electron would thus be in one of the states we have drawn in Fig. 4.3 of §4.3.2, provided that the quantum angular momenta aligned with the external magnetic field. Stern thought they would. In that case the beam of silver atoms would become a beam of magnets with magnetic moments that should be $\pm\boldsymbol{\mu}_{\mathrm{B}}$. One would thus expect a statistical distribution of silver atoms with the magnetic moments $\pm\mu_{\mathrm{B}}$ in the beam. If he could separate the atoms according to their magnetic moment, he would have detected the spatial quantization to which Pauli had referred.

Remarkably, perhaps, Gerlach was already working on the separation of bismuth atoms in an inhomogeneous magnetic field and was gaining valuable experience. Stern, however, was Born's assistant. He had first to obtain Born's blessing before he could go to Gerlach and discuss a new focus for his expertise. So his first objective that morning was to convince Born of the value of his idea [117].

According to his recollection, Born was not so enthusiastic. Born was a theoretician. He understood the theoretical ideas behind the Sommerfeld–Debye model, but to him these were mathematical constructs, while to Stern they were real physical concepts to be observed. Born tried to dissuade Stern, but Stern said it was worth a try and eventually received Born's blessing [117].

The fact that Stern was already successfully pursuing some of the fundamental questions of gas theory with beams of silver atoms had already impressed Born. Because of the limited physical space in Frankfort, Stern was conducting experiments in the same room as Born [117].

Then Stern went to see Gerlach. Although we do not have the details, we may imagine that Stern found Gerlach working on his apparatus for the bismuth beam experiments and would have first asked several pointed questions about the experiment and the apparatus. Stern would have mentioned that something else could be done with the spatially inhomogeneous magnetic field and asked Gerlach if he knew about spatially dependent quantization. Gerlach would have confessed that he did not. Then, after some rather compressed explanation, Stern might have asked, "Shall we do it?" And we know that the response from Gerlach was affirmative [117].

Figure 5.18 is a schematic drawing of the basic apparatus used by Stern and Gerlach. Like many fundamental experiments in physics, the design of the basic apparatus was simple. A beam of silver atoms, produced from an oven like the one Stern had been using (operating at 1000 °C), was collimated into a very thin beam (by two narrow slits 0.03 mm wide) and directed between the magnetic poles of Gerlach's specially designed magnet, producing a strong, inhomogeneous field (3.5 cm long with field strength about 0.1 T and gradient 10 T/cm). The silver atoms in the collimated beam would then separate vertically depending on the value of the projection quantum number of each atom. The beam leaving the magnetic poles

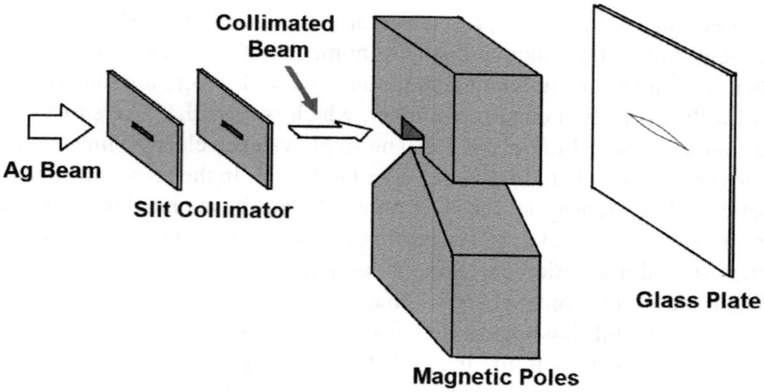

Fig. 5.18 Stern–Gerlach apparatus. A beam of Ag atoms produced by an oven enters from the left and passes through a double slit collimator, which produces a very thin beam. This beam enters the inhomogeneous magnetic field between the magnetic poles. The beam is separated by the inhomogeneous field and condenses on the cold glass plate ([118] Drawn by CSH)

was subsequently directed onto a cold glass plate where the atoms would condense. Stern's calculations based on the properties of the silver atoms and the attainable magnetic field intensity indicated that the experiment was barely feasible. The beam was also of very low intensity. So the experimental measurement time would need to be long. However, the duration was limited to a few hours due to technical limitations. The splitting of the beam at the collector was 0.2 mm, which was 6.7 times the beam width and so could be detected provided enough atoms had condensed on the glass plate. However, a misalignment of the collimating slits or the magnet of greater than 0.01 mm was sufficient to spoil the data from an experimental run [118].

The data collected as silver atoms deposited on a glass plate were not easily seen. Stern related a story after an early run in which Gerlach removed the detection plate and could see nothing. He then handed the plate to Stern, who happened to be smoking a cheap cigar which contained sulfur in the leaves. Stern explained that he could not afford good cigars. As he looked at the plate, with Gerlach looking over his shoulder, the data began to emerge as a black image on the plate. The sulfur in the cigar smoke had formed silver sulfide which is black. According to Stern, it was like watching a photograph emerge in the darkroom. After this episode the pair turned to a photographic method for gathering data. The cigar smoking, nevertheless, continued [118].

However, the experiments did not yield definitive results. Months went by with no conclusive results. Stern's confidence was slipping at times. And Gerlach had an exchange with Debye in which Debye had pointed out that the spatial orientation of atoms was not something that should be considered as real. Funding in Germany had also become very difficult as the economy slid downhill. And then in 1921 Stern was offered a post as professor of theoretical physics at the University of Rostock, which he accepted.

Stern and Gerlach were able to meet in Göttingen early in 1922 to consider the situation they faced and decided to end the work. But there was a railroad strike and Gerlach couldn't return immediately to Frankfort. This gave him a day to review the situation, going over all the details. He finally decided that he would continue, beginning with an improved alignment of the apparatus.

With this new beginning, he achieved a clear splitting of the collimated beam. Without the magnetic field there was a single line on the film and with the magnetic field that single line opened indicating a splitting of the beam. He then sent a telegram to Stern with the sentence "Bohr is right after all." Stern later recalled that his own excitement at receiving this telegram was overwhelming [243]. Gerlach also sent a postcard to Bohr containing a picture of the data with and without the magnetic field [118].

We will leave this remarkable story here in 1922, with its triumph, humor, and near tragedy. At the time of this success in the laboratory the quantum mechanical spin of the electron was not yet known. Nor was there yet a real quantum theory. We have been able to see great theoreticians working to construct something new from the old quantum theory. And we have seen great experimentalists spending days and nights laboring over their apparatus and finally coaxing out results from equipment that failed them more often than not.

5.6 Summary

In this chapter we considered experiments related to important ideas that emerged from the beginnings of the quantum theory.

The first of these was the Bohr concept of the stationary orbit. Bohr's ideas led to the prediction that the observed frequency of emitted radiation from an atom was a function of its nuclear charge. This was supported in the He^+ spectra from stars. Moseley extended this idea to the emission of X-radiation and found the relation between atomic number and nuclear charge experimentally. Then Franck and Hertz identified a transition in Hg, helped by Bohr's insights, to demonstrate the validity of the Bohr atomic model.

The Einstein photon presented great difficulties for many scientists because it simply stood at cross purposes to one of the most beautiful of physical theories: Maxwell's electrodynamics. But a careful reading of Einstein's 1905 paper shows that this idea was solidly based in the physics, even though it was radical. Millikan's resistance to Einstein's photon remained, although his experimental results did not seem to support him, while Compton's experiment, with some false starts on the theoretical interpretation, left no doubt.

Bohr's resistance to the Einstein photon and the BKS paper are historically factual, even though we often explain the Bohr 1913 paper in terms of emission and absorption of photons. It is, perhaps, historically fortunate that the answers came so quickly, and from almost supreme experimental ingenuity, based on the Compton effect.

The Stern–Gerlach experiment could almost occupy a place all by itself. It was based on Stern's taking spatial quantization seriously, even though theoreticians such as Born and Debye considered them to be only mathematical constructs. The experiments were also very difficult to perform, even though the idea seems so simple. The final push by Gerlach represents a high point in the patience and genius of the experimentalist.

There is more here to see than just experimental verification. There is a link between theory and experiment to form our understanding of what seems to be emerging. In our quote at the beginning of this chapter we see that Einstein understood the connection in a way that some may not have fully realized. Stern and Gerlach then provided experimental support for this idea.

Truth cannot be subdivided.

Chapter 6
De Broglie's Particle Wave

Thus with every advance in our scientific knowledge new elements come up, often forcing us to recast our entire picture of physical reality.

Louis de Broglie

6.1 Introduction

Prince Louis Victor Pierre Raymont de Broglie (1892–1987), who is normally referred to simply as Louis de Broglie, actually had an inherited title, which he obtained after the death of his elder brother Maurice (1875–1960), as the 6th duc de Broglie. The de Broglies were an important family in France and for generations served France as soldiers, politicians, and diplomats. Four de Broglies were Marshals of France. However, the French Revolution was not an easy time for the family, and at least one died on the guillotine [34, p. 262].

On the death of their father in 1906, Maurice assumed responsibility for his younger brother Louis' education. Although Maurice obtained the degree *Docteur des Sciences* in 1908, he did not initially encourage Louis to study science. Rather he thought Louis was best suited for politics and diplomacy. However, in 1911 Maurice served as one of the secretaries to the first Solvay Conference. Along with Paul Langevin (1872–1946), he was responsible for preparing notes from the discussions to be published in the proceedings of the conference. Louis was able to read the notes and became enthused by the problems in the quantum and relativity theories that were being discussed. He then switched to the study of sciences. In 1913 he received his *Licence des Sciences* at the Sorbonne University in Paris and elected to fulfill his military obligation. The following year, World War 1 began and Louis and Maurice were fully occupied with that until 1919 [188, p. 549] (Fig. 6.1).

C. S. Helrich, *The Quantum Theory—Origins and Ideas*, History of Physics,
https://doi.org/10.1007/978-3-030-79268-8_6

Fig. 6.1 Chapel of the main Sorbonne building (Place de la Sorbonne - Paris V). Université de Paris (Photographer Mbzt)

Maurice de Broglie was a retired naval officer who, during the war, invented a radio receiver for submarines. Louis was assigned to a radio transmitting station at the Eiffel Tower, a short distance from where Léon Brillouin (1889–1969) was located. Brillouin conducted some experiments with Louis and was involved with Maurice in the mounting and testing of the submarine receiver [188, p. 549, footnote 7].

After the war, Louis worked with Maurice in his private laboratory set up to study X-rays. As a result he was able to publish some short papers and became involved in long discussions with Maurice on the true nature of X-rays . These discussions, Louis said, made him reflect deeply on the continual need to connect the wave and particle aspects of these rays. His time in Maurice's laboratory also gave Louis a chance to publish some of his formative thoughts on Einstein's photons, which he called light particles. He added some ideas about the origins of interference using light particles [188, pp. 550–553].

Louis also had discussions with Léon Brillouin regarding matter waves. Once, with Brillouin, he considered Rutherford's observation that α and β particles were bent in opposite directions in a magnetic field, while γ rays were not affected. According to Brillouin, de Broglie said that these must all be very similar: they had to be either all waves or all particles. Brillouin relates that de Broglie had to think and wonder a long time before he finally chanced upon the idea that a light particle with energy $h\nu$ would have momentum $h\nu/c$. It finally appeared as a single sentence in a paper de Broglie wrote in January of 1922. Brillouin believed that the origin of de Broglie's phase wave idea lay in this remark [188, p. 553].

The final concept of matter waves crystallized for de Broglie in the summer of 1923. He said that the long discussions he had had with his brother always came to the same point: X-rays had to be both waves and particles at one and the same time. At some point during that summer he came to the conviction that this duality had to extend to material particles. Specifically, electrons had to have this dual nature. And he recalled the Hamilton–Jacobi equation of classical mechanics. This equation represents the most elegant formulation of the mechanics of particles, but it can also be used to treat geometrical optics, and in particular Fermat's principle for light rays, which can be formulated as a generalization of Hamilton's variational principle. During that summer de Broglie wrote three notes for the journal *Comptes rendus*, which later appeared in his thesis [72, 188, p. 554].

6.2 De Broglie's Thesis

The core idea that de Broglie put forward in his 1923 doctoral thesis was that a material body in motion could always be considered as a wave phenomenon. He could not base this simply on the long discussions with his brother or even those with Brillouin. Einstein's development of the photon, that we considered in Sect. 3.3 was instrumental in his thinking. But this was only an addition to the ideas he may already have had on X-rays. He now had a theoretical formulation of his fundamental idea, with which he could begin his thesis.

He introduced his thesis with a survey of the history of physics from the 16th through to the 20th century, including an analysis of Einstein's ideas on light, but he began the first chapter by introducing the phase wave, which was central to the formal development of his idea. In what follows we shall present the ideas that de Broglie developed in the first chapter of his thesis. To clarify them, we shall deviate from his presentation at times. Our designation of subsections follows his, however.

6.3 The Phase Wave

6.3.1 *The Relation Between the Quantum and Relativity Theories*

First de Broglie recalled the equivalence of matter and energy that is expressed as

$$energy = mass \times c^2. \tag{6.1}$$

The principle of inertia tells us that a body at rest has an inertial mass of m_0 and a *proper* energy of m_0c^2. If that body is in motion with a velocity of $v = \beta c$, the mass of the body is then $m_0/\sqrt{1-\beta^2}$ and the energy is $m_0c^2/\sqrt{1-\beta^2}$.

In order to introduce quanta into relativistic dynamics, it seemed to de Broglie that the main idea in the quantum theory was the impossibility of considering an amount of energy without attributing a frequency ν to it according to

$$energy = h\nu, \tag{6.2}$$

where h is Planck's constant of action. This he called the *quantum relationship*. Although he considered the *action* to be a very abstract notion, he noted that it seemed to play an important role in the quantum theory. However, after considerable reflection on light quanta and the photoelectric effect, he decided to focus on energy rather than action.

He struck upon the following meta-law: to each proper mass we may associate a frequency ν_0 by equating $h\nu_0$ to the proper energy as

$$h\nu_0 = m_0 c^2, \tag{6.3}$$

where the frequency ν_0 was to be measured in the rest frame of what he termed the *energy packet* or *parcel*. Equation (6.3), he wrote, would be the basis of the theory he was developing. He admitted that, as with any hypothesis, it would only be as good as the consequences that could be deduced from it.

He also chose to refer to the limiting speed c as the "limit speed of energy" (quotations in original), for reasons which he promised to explain. He was in fact carefully developing an idea here to which he had come during those long discussions with his brother and with Brillouin.

He asked whether we should consider the periodic motion associated with ν_0, as defined by (6.3), to be a motion in the interior of the energy parcel. The electron could be considered the archetype of an isolated parcel of energy or matter, but he pointed out that the electron would exist over the whole of space and would not be isolated.[1]

De Broglie then considered the frequency of the motion associated with a moving parcel of energy. This parcel of energy could be studied by someone moving with the parcel or someone stationary who was observing the moving parcel. The time associated with the periodic motion would appear to be slowed down by the motion of the parcel.[2]

The stationary observer would measure a frequency ν_1 for the moving parcel, which, using (6.3), is

---

[1] This was an issue he said he would develop, but offered no proof of it.

[2] The time associated with a wave is the period. This is a time interval. The frequency is the reciprocal of the period. If an observer moving with the wave measures the wave period to be τ' and a stationary observer measures the period of the moving wave to be τ, then

$$\tau' = \tau\sqrt{1-\beta^2},$$

where $\beta = v/c$ with v the speed of the wave. The frequencies are then related by

$$\nu = \nu'\sqrt{1-\beta^2}.$$

$$\nu_1 = \nu_0\sqrt{1-\beta^2} = \frac{m_0c^2}{h}\sqrt{1-\beta^2}. \tag{6.4}$$

Since the parcel is moving, the stationary observer would find the mass to be $m_0c^2/\sqrt{1-\beta^2}$ and would associate with it a frequency

$$\nu = \frac{1}{h}\frac{m_0c^2}{\sqrt{1-\beta^2}}. \tag{6.5}$$

The frequencies ν_1 and ν are different. This fact, de Broglie wrote, had long intrigued him, and it was what led him to the *theorem of phase harmony*:

> *A periodic phenomenon as seen by a stationary observer has a frequency $\nu_1 = h^{-1}m_0c^2\sqrt{1-\beta^2}$ that appears to be constantly in phase with a wave of frequency $\nu = h^{-1}m_0c^2/\sqrt{1-\beta^2}$ propagating in the same direction with a velocity $V = c/\beta$.*[3]

To establish this de Broglie assumed that there was phase harmony initially, at time $t = 0$ for each observer. He also assumed that the periodic phenomenon, which he identified as a *moving object*, had moved a distance $x = \beta ct$ in the time t. The stationary observer would then record the phase of the wave with frequency ν_1 at the time t as

$$\nu_1 t = \frac{m_0c^2}{h}\sqrt{1-\beta^2}\left(\frac{x}{\beta c}\right). \tag{6.6}$$

Taking the time recorded by an observer moving with the object (the *proper time* of relativity) to be t_0, then from the Lorentz transformation (see [139, p. 280])

$$t_0 = \frac{1}{\sqrt{1-\beta^2}}\left(t - \frac{\beta x}{c}\right). \tag{6.7}$$

The phase for this observer moving with the object is then $\nu_0 t_0$, which, with (6.3), is

$$\nu_0 t_0 = \frac{m_0c^2}{h}\frac{1}{\sqrt{1-\beta^2}}\left(t - \frac{\beta x}{c}\right). \tag{6.8}$$

Since $t = x/\beta c$, (6.8) becomes

$$\nu_0 t_0 = \frac{m_0c^2}{h}\sqrt{1-\beta^2}\left(\frac{x}{\beta c}\right). \tag{6.9}$$

We see that (6.9) and (6.6) are identical. This equality establishes de Broglie's theorem of phase harmony.

---

[3] At this point de Broglie gave no reason for this particular velocity. He provided his reasoning later, using the relativistic geometry developed by Hermann Minkowski.

There remains the remark at the end of the statement of the theorem of phase harmony that the velocity of the wave is $V = c/\beta$. Because $\beta < 1$, the wave we are considering has a velocity greater than the velocity of light. De Broglie notes this and accepts that no energy can be carried by this wave. Since the theorem also deals with the phase of the wave, he points out this is therefore a *phase wave*.

6.3.2 Phase and Group Velocities

The concept of phase wave is something we encounter in a Fourier representation of a propagating electromagnetic pulse. A very short burst of electromagnetic energy arising, for example, from a flash lamp is represented by a group of phase waves with a spread of frequencies and a corresponding spread of velocities [139, pp. 250–254]. De Broglie had already claimed that the electron might be thought of as an energy parcel and had spoken of a wave associated with this parcel, but he did not identify the wave as the parcel.

In his thesis, de Broglie concluded that the velocity of the phase wave was $V = c/\beta$, using the space-time geometry of Hermann Minkowski (1864–1909) [197, pp. 75–91]. We shall not follow the geometrical space-time approach here. Rather we shall obtain the phase wave velocity as part of a general description of the phase wave representation of an energy parcel, as this follows de Broglie's approach to the topic.

We consider the parcel of energy (the electron) to be represented by a group of phase waves with frequencies in the neighborhood of ν, that is $\nu \to \nu + \delta\nu$, and velocities in the neighborhood of V, that is $V \to V + \delta V$. Specifically, considering two such waves, their sum is

$$\sin\left[2\pi\left(\nu t - \frac{\nu}{V}x + \phi\right)\right] + \sin\left[2\pi\left(\nu' t - \frac{\nu'}{V'}x + \phi'\right)\right]$$
$$= 2\cos\left[2\pi\left(\frac{\delta\nu}{2}t - x\frac{\mathrm{d}(\nu/V)}{\mathrm{d}\nu}\frac{\delta\nu}{2} + \psi'\right)\right]\sin\left(2\pi\left(\nu t - \frac{\nu}{V}x + \psi\right)\right). \quad (6.10)$$

The sum of the two slightly separated sine waves then produces a sinusoid modulated by a cosine wave at the frequency $\delta\nu/2$. The velocity of this cosine wave is denoted by U, where

$$\frac{1}{U} = \frac{\mathrm{d}(\nu/V)}{\mathrm{d}\nu} = \frac{\mathrm{d}(\nu/V)/\mathrm{d}\beta}{\mathrm{d}\nu/\mathrm{d}\beta}. \quad (6.11)$$

This U is the group velocity of the modulated wave. It is the velocity at which energy is transported, and it is thus the velocity at which the energy parcel (the electron) moves (cf. [139, 278, p. 385]).

To calculate the group velocity, we begin with ν from (6.5) to obtain

$$\frac{\nu}{V} = \left(\frac{m_0c^2}{h}\right)\left(1-\beta^2\right)^{-1/2}\frac{1}{V}. \tag{6.12}$$

Then,

$$\frac{\mathrm{d}\left(\nu/V\right)}{\mathrm{d}\beta} = \left(\frac{m_0c^2}{h}\right)\left(1-\beta^2\right)^{-1/2}\left[\beta\left(1-\beta^2\right)^{-1}\frac{1}{V} - \left(\frac{1}{V^2}\right)\frac{\mathrm{d}V}{\mathrm{d}\beta}\right]. \tag{6.13}$$

And from (6.5), we have

$$\frac{\mathrm{d}\left(\nu\right)}{\mathrm{d}\beta} = \left(\frac{m_0c^2}{h}\right)\beta\left(1-\beta^2\right)^{-3/2}. \tag{6.14}$$

Therefore, from (6.13) and (6.14), we have

$$\frac{1}{U} = \frac{\mathrm{d}\left(\nu/V\right)/\mathrm{d}\beta}{\mathrm{d}\left(\nu\right)/\mathrm{d}\beta} = \frac{1}{\beta V}\left[\beta - \left(\frac{1}{V}\right)\left(1-\beta^2\right)\frac{\mathrm{d}V}{\mathrm{d}\beta}\right]. \tag{6.15}$$

But we have already defined $\beta = v/c$, where v is the velocity of the energy parcel. In our present development, this is the group velocity U. Hence, $U = \beta c$.

6.3.3 Phase Waves in Space-Time

This is the final section in the first chapter of de Broglie's thesis. In this section he turns to Minkowski's space-time geometry to prove that $V = c/\beta$. His proof is geometrical in the two-dimensional space (x, ct). Our proof here will avoid the extra space that would be required to introduce Minkowski's geometry.

Substituting the value $U = \beta c$ for the group velocity into (6.15), we have

$$\frac{V^2}{c} - \beta V + \left(1-\beta^2\right)\frac{\mathrm{d}V}{\mathrm{d}\beta} = 0, \tag{6.16}$$

which is a nonlinear differential equation for V. We choose not to attempt a solution to this equation. However, we can simply insert the proposed solution $V = c/\beta$ to show that it is valid. We thus realize that the two waves in the first line of (6.10) are phase waves from which the energy parcel, moving at the group velocity $U = \beta c$, has been constructed.

The phase velocity V of these phase waves is the product of the frequency ν and wavelength λ of the waves. With ν from (6.5), we then have

$$V = \left(\frac{m_0c^2}{h}\right)\left(1-\beta^2\right)^{-1/2}\lambda. \tag{6.17}$$

With $V = c/\beta$, (6.17) becomes

$$\frac{m_0\lambda}{\left(1-\beta^2\right)^{1/2}} = \frac{h}{\beta c}. \tag{6.18}$$

The velocity of the energy parcel is $v = \beta c$ and the mass of the moving parcel is

$$m = \frac{m_0}{\left(1-\beta^2\right)^{1/2}}. \tag{6.19}$$

Therefore, with the momentum $p = mv$ for the energy parcel,[4] (6.18) becomes

$$\lambda = \frac{h}{p}. \tag{6.20}$$

This is the de Broglie wavelength associated with a moving particle (the energy parcel).

In Chap. 1 of de Broglie's thesis, the equation

$$\nu = \frac{m_0 c^2}{\left(1-\beta^2\right)^{1/2}} \tag{6.21}$$

appears at the end of the discussion on relativity. With $\nu\lambda = V = c/\beta$, equation (6.21) becomes (6.18). However, de Broglie did not write this equation in the form (6.20). This may seem curious, since we often consider de Broglie's main contribution to be the result expressed in equation (6.20).

6.4 Connections

After the development of his concept of the phase wave in the first chapter, the remaining chapters of de Broglie's thesis are essentially a reflection on this idea, in light of the fundamental principles of theoretical physics as understood in 1923.

In Chap. 2, de Broglie outlined the variational principles due to Maupertuis, Hamilton, and Jacobi, that formed analytical mechanics, and the principle of refraction of light due to Pierre de Fermat (1607–1665). Jacobi's time-independent form of the Hamilton–Jacobi equation appears in Chap. 3 of the thesis. Also in Chap. 3, de Broglie addressed the quantum stability conditions for particle trajectories. There he obtained the stability conditions for the Bohr atom, arriving at the condition that the angular momentum of the electron in a stable Bohr orbit must be a multiple of $h/2\pi$.

[4] The expression $p = mv$ is also relativistically correct provided m is the relativistic mass.

In the final chapter of his thesis (Chap. 7), de Broglie discussed the issue of statistical mechanics. He referred to it as quantum statistical mechanics, although it was not what we presently know as quantum statistical mechanics, since the physics of the day was simply not yet ready for that.

6.4.1 Defense of the Thesis

When de Broglie defended his thesis, his arguments left a good impression on the committee, although not all its members were convinced of the validity of the ideas. Langevin was his principal thesis supervisor and supported de Broglie and his ideas. Of course, Langevin had spoken with Einstein about de Broglie's ideas before the date of the examination. Einstein's positive and indeed friendly interest in de Broglie's ideas convinced Langevin to take a positive position. Einstein asked Langevin if he had a spare copy of the thesis. Fortunately de Broglie had had three copies made, so one could be sent to Einstein [188, p. 569].

6.4.2 Influences

The next major step was taken by Erwin Schrödinger (1887–1961), who began with the time-independent form of the Hamilton–Jacobi equation. In the third section of his first paper on wave mechanics, Schrödinger noted explicitly the inspiration he had gained from de Broglie's thesis. He commented specifically on the phase wave idea that de Broglie had developed. Then he pointed out that the principal difference between the two approaches was in the fact that de Broglie had developed a picture of travelling waves, while he would consider time-independent eigensolutions to a wave equation [249, pp. 372–373].

6.5 Davisson–Germer Experiment

Clinton J. Davisson (1881–1958) was frail of frame throughout his life. He graduated from high school at age 20 and received a one year scholarship to the University of Chicago, where he spent six years, since his studies were interrupted by occasional lack of funds. At Chicago, he found physics "concise and orderly," and was inspired by Millikan. Before finishing his degree at Chicago, he was a part-time instructor in physics at Princeton University, where he came under the influence of Owen Richardson (1879–1959). He earned his Ph.D. under Richardson at Princeton in 1911 and subsequently married Richardson's sister, Charlotte, whom he had met when she was visiting from England in 1911. After the honeymoon, Davisson joined the Carnegie Institute of Technology as an instructor. But the 18 hour teaching load left him no time for research, except for the summer he spent at the Cavendish Laboratory with J.J. Thomson in 1913. Richardson returned to England in 1914 to become Wheatstone Professor of Physics at King's College London.

In 1917 Davisson was unable to enlist in the U.S. Army because of his physical frailty. He took a leave of absence from Carnegie Tech to do war-related work at Western Electric, which was the engineering arm of the American Telephone and Telegraph Company (AT&T). This would later become Bell Telephone Laboratories. After the war he turned down a promotion at Carnegie Tech and accepted a permanent position at Western Electric [121].

Bell Telephone Laboratories, then in New York City, was an industrial research center. The rapid growth and capabilities of the industrial laboratory in the post World War 1 period was one of the key features of the developing American scientific system. The mission of an industrial laboratory differed from that of an academic or national laboratory in that the projects undertaken were directly related to the perceived needs of the company supporting the laboratory.[5] The laboratory director could, however, use his own judgement to provide the freedom necessary to pursue any investigation that seemed critical for scientific reasons, but might not be obviously directly related to any industrial question. At this time, the director of the research section at Bell Laboratories was Harold D. Arnold (1883–1933). Figure 6.2 is a photo of the Bell Telephone Laboratories building in New York City, taken in 1936.

In March of 1917 Davisson was assigned an assistant, Lester H. Germer (1896–1971), who had just graduated from Cornell. Two months later, however, Germer volunteered for service in the army and became a pilot in the aviation section of the Signal Corps [121]. After the war, following three weeks of rest, Germer returned to Bell Laboratories to work with Davisson. They were assigned the investigation of thermionic emission in oxide-coated cathodes under positive ion bombardment. This led to an investigation of the nature of secondary emission from grids and plates. According to Germer's later recollection, this study originated because of a patent dispute between Arnold of Western Electric and Irving Langmuir (1881–1957) of General Electric. The primary problem in the lawsuit, however, dealt with the understanding of the physics prior to 1913. Therefore, data from 1920 were of little value [240, pp. 125, 130].

In 1920, Bell Laboratories initiated a cooperative program for advanced degrees with Columbia University, which allowed those chosen to freely elect their course of study and curtail their work time. Germer was one of the first to take advantage of this, obtaining an M.A. in 1922 [240, p. 120]. At that time Davisson was interested in pursuing a sideline investigation of secondary electron emission under electron bombardment. This sideline was encouraged by Arnold, who supplied Davisson with a new assistant, Charles Kunsman (1890–1970). Kunsman had just received a Ph.D. from the University of California at Berkeley.

Davisson and Kunsman found that 1% of the electrons were scattered back toward the electron gun with virtually no energy loss. The beam energy was 150 eV. The

[5] There was a clear difference between the mission of Bell Telephone Laboratories and the PTR in Berlin-Charlottenburg, Germany.

Fig. 6.2 Bell Telephone Laboratories Building in New York City 1936. (Public Domain. Author unknown)

scattered electrons showed two maxima. One of these was in the direction of the incoming beam and the other was at an angle. This angle was affected by the energy of the incoming beam. Davisson was profoundly impressed by this result. He noted the similarity of the electron scattering to the scattering of alpha particles that had been engaging Rutherford and his group in Manchester. In Davisson's own words [121]:

> What we were attempting [...] were atomic explorations similar to those of Sir Ernest Rutherford [...] in which the probe should be an electron instead of an alpha particle.

In 1921, Davisson and Kunsman submitted a two-column paper to *Science*. But the subsequent experiments yielded unimpressive results and they were not published. Kunsman left the company at the end of 1923 and Davisson abandoned the scattering investigations [66, 121, 240, pp. 131–132].

Germer had fallen ill, missing 15 months at work as a result. However, in October of 1924, he was put back on the project replacing Kunsman [121]. Preliminary experiments were carried out to cross-check the new data with the data of the previous experiments carried out by Davisson and Kunsman. Then on 5 February 1925, the now famous accident occurred. A liquid air bottle exploded while the target was at a high temperature. The tube used for the experiment was broken and the target oxidized. This was not the first accident that had occurred in the course of these experiments, but this time the target was not replaced. The oxide was removed by vaporization along with a layer of the target, after prolonged annealing in H_2 and in vacuum [121, 240, p. 139].

On 6 April the experiments began again. At first the results were not different than those previously obtained. But then on 12 and 14 May, rather than the single offset peak previously observed, there were multiple peaks with varying intensities. To understand what might have happened Davisson and Germer cut open the apparatus and, with the help of the microscopist Francis F. Lucas (1885–1961), examined the structure of the target. Instead of the many microscopic crystals present previously, they observed only a few large crystals. The results, therefore, came from the scattering of electrons by crystals rather than by atoms. Fortunately, there was a group at the laboratory doing research on crystal structures [121, 240, p. 140].

Davisson decided that they should now consider scattering from a single crystal with known orientation with respect to the electron beam. Because of the demands of other experiments in the laboratories, a single crystal was not available until April 1926, provided by the company's metallurgist Howard Reeve. They mounted the crystal so that both the polar and azimuthal angles could be changed. The results were disappointing. There was no variation in the reflected beam from variation in the polar angle, and only meagre dependence on the azimuth [121, 240, p. 141].

Arturo Russo points out that electron diffraction might never have been discovered at Bell Laboratories had Davisson not needed a rest and returned to England with his wife for a second honeymoon during the summer of 1926. According to Charlotte Davisson, they were fortunate that her sister and brother-in-law at Princeton University were able to care for the children [121]. While in England, (Clinton) Davisson attended the Oxford meeting of the British Society for the Advancement of Science. There he heard a lecture by Born on the new quantum theory and the wave nature of matter proposed by de Broglie [72, 73, 240, p. 141].

Davisson had previously contacted Born for advice on crystals, since he was a recognized expert. At the time, Born had suggested to Davisson that the electron scattering data might be the result of forces from different planes in the crystal. This first contact had left little impression on Born until he discussed the issue later with Franck, who raised the possibility of electron waves as proposed by de Broglie.

Born and Franck realized that the data from Bell Laboratories might indicate diffraction of electron waves. Together they decided that the issue was worth pursuing and assigned a new graduate student in Franck's laboratory, Walter Elsasser (1904–1991), to the problem. In July of 1925 Elsasser was ready to suggest that the 1921 results of Davisson and Kunsman might be indicative of the diffraction of electron waves. He was not, however, in a position to conduct any experiments of his own [240, p. 143].

In his lecture at Oxford, Born cited de Broglie's work and Schrödinger's papers on wave mechanics. Then he noted that the experiments by Davisson and Kunsman in 1921 had provided evidence for the diffraction of electron waves. This resulted in an extensive discussion after the lecture with Born, Franck, and Douglas Hartree (1897–1958) providing Davisson with an introduction to Schrödinger's papers, which dealt extensively and directly with matter waves, de Broglie's papers, and the new matrix mechanics due to Born, Heisenberg, and Jordan. Davisson had normally kept

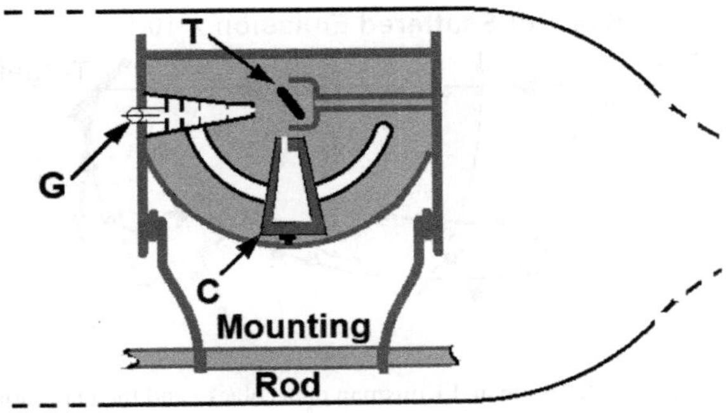

Fig. 6.3 Davisson and Germer's apparatus in 1927. **T** is the Nickel target, **G** is the electron gun, and **C** is the double Faraday box collector for the scattered electrons. The electrons entering the Faraday box are registered by a sensitive galvanometer. This apparatus was sealed inside a glass tube, which is indicated partially. The angular location of the Faraday box was adjusted by rotating the tube about an axis perpendicular to the drawing. (Simplified from Fig. 2 of [70]. Drawn by CSH)

himself informed of developments in physics, but it seems he was not yet aware of these advances in quantum mechanics [121].

Davisson borrowed reprints of the Schrödinger papers and a German–English dictionary from Richardson to study on the return trip to New York. When he was back at Bell Laboratories he had an idea for the experiment he and Germer should perform. Davisson wrote to Richardson [121]:

> I am still working at Schrödinger and others and believe that I am beginning to get some idea of what it is all about. In particular I think that I know the sort of experiment we should make with our scattering apparatus to test the theory.

Arnold acknowledged the importance of the idea and assigned the mechanical engineer C. Calbick as an assistant [121].

The experiments began in December of 1926 [240, pp. 144, 145]. The first results in accordance with theory were obtained on 6 January 1927 and the first announcement was sent to *Nature*[6] [67]. The results were also presented at the *American Physical Society* meeting in Washington in April [69] and a *Bell Laboratories Record* report was published in April [68]. A lengthy review paper then appeared in the *Physical Review* in December of 1927 [70, 121].

Figure 6.3 is a simplified drawing of the main parts of the 1927 apparatus. As in all previous experiments, the apparatus was mounted inside a glass tube containing a high vacuum to prevent absorption of the electrons. An approximate outline of this tube is provided in Fig. 6.3 from pictures of the apparatus in the National Museum of American History at the Smithsonian Institution in Washington, DC. Figures 6.2 and 6.4 in Davisson and Germer's *Physical Review* paper are detailed technical drawings of the apparatus from each side.

[6] Gehrenbeck presents an experiment by experiment account of this period [121].

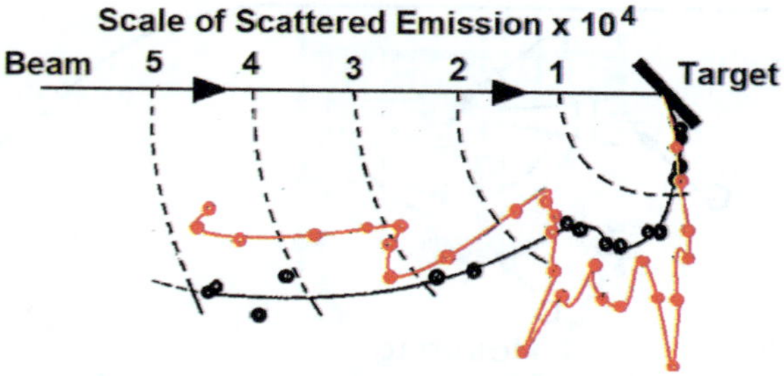

Fig. 6.4 Composite of the Davisson and Kunsman data (black) and the Davisson and Germer data (red). Data plotted in black are from Fig. 1 of the November 1921 *Science* article. Data plotted in red are from Fig. 1 of the December 1927 *Physical Review* article. (Public Domain. Plotted by CSH)

The electron gun **G** in Fig. 6.3 provided a collimated beam of electrons directed at the nickel target **T**, which could be rotated to position selected crystal lattice planes relative to the electron beam. The collector **C** was a double Faraday box with output to a sensitive galvanometer, and was similar to the collector used by Millikan (see Fig. 5.8). The angle of the collector relative to the beam of electrons was selected by rotating the tube and measured on a scale on the side opposite the one shown in the drawing.

Figure 6.4 shows a composite of the data from Davisson and Kunsman's paper in *Science* which attracted the attention of Born's group, and the data from the Davisson and Germer paper in the *Physical Review*, which was their final statement of 1927. The plot in Fig. 6.4 is configured so that the beam is aligned with the one in Fig. 6.3. The scale is correct for the 1921 (black) data. The scale of the 1927 (red) data has been adjusted to make the difference between the two results clear.

The plot is polar with an origin at the point at which the incoming electron beam strikes the target. A polar angle of zero is defined by the negative direction of the incoming electron beam. Positive diffracted beam angles are measured in a clockwise direction from this. The value of the radial component is the ratio of the diffracted electron intensity at the corresponding angle to the intensity of the incoming beam. The scale in our drawing, as in Davisson and Germer's plot, has a factor of 10^4 as shown in Fig. 6.4. This is an indication of the experimental expertise.

The relative minima and maxima of the 1921 data are at polar angles of 65° and 75°, respectively. Davisson and Kunsman indicated these angles in their paper. We have removed them to avoid clutter. Knowing that the relative maximum, so crucial in attracting the attention of Born's group, was at 75° will help the reader to understand the polar plot.

The black data points in Fig. 6.4 have been taken directly from Fig. 6.1 in the *Science* paper. The red points are those at 5° intervals read from the *Physical Review* paper. There were many more data points in the graphs in the 1927 paper in order to clearly indicate the structure that can be seen in Fig. 6.4.

6.6 Summary

As we have seen in this chapter, de Broglie developed his ideas slowly and carefully before setting them in mathematical form. He was ultimately guided by both the mathematics and the physics, and his knowledge of the physics was extensive, as is clear from his thesis.

His theorem of phase harmony stands at the beginning of his idea of the phase wave representation of the energy parcel, and he arrived at this theorem through a relationship between frequencies that he considered central to the quantum theory. These were bold and inspired steps, but it was not the result of a sudden inspiration.

He subsequently reflected on this in terms of what was known in 1923. Of particular importance was the connection between Hamilton's and Fermat's principles. De Broglie wanted to give his ideas more substance and make them more easily presentable. Here was a mathematical connection between the particle trajectory and the geometrical treatment of light. But de Broglie pushed it farther, concluding that the particle momentum had to be related to the wavelength of the phase wave. The importance of this relationship is primary, since it provides something that can be measured in the laboratory.

In our last section we picked up the story of the measurement, which was the perfectly serendipitous. That story involved the leading American industrial laboratory of the day and also acquainted us with Schrödinger.

Chapter 7
Göttingen Quantum Theory

> Almost all progress in science has been paid for by a sacrifice; for almost every new intellectual achievement, previous positions and conceptions had to be given up. Thus, in a way, increasing knowledge and insight continually diminishes the scientist's claim to "understand" nature.
>
> Werner Heisenberg

> Mathematics, as an expression of the human mind, reflects the active will, the contemplative reason, and the desire for aesthetic perfection.
>
> Richard Courant

7.1 Introduction

The mathematical structure of a theory of quantum mechanics based on matrices was first developed in the work of Werner Heisenberg (1901–1976) in the early summer of 1925 [135], then picked up by Born and Pascual Jordan (1902–1980) [23] by the end of the summer [23], and essentially completed by Born, Heisenberg, and Jordan together in Göttingen in the fall [24]. The manuscript by Born, Heisenberg, and Jordan was received by *Zeitschrift für Physik* on 16 November 1925.

Heisenberg received his doctorate under Sommerfeld at the LMU in Munich in 1923 and then his habilitation under Born at the Georg-August University in Göttingen in 1924 [131]. Born received his doctorate under Felix Klein in Göttingen in 1907. Jordan received his doctorate under Born in Göttingen in 1924. Jordan had an unusually wide range of interests and finally settled on theoretical physics after some exploration ([187], pp. 44–59). The great mathematician Hermann Minkowski (1864–1909) was also in Göttingen from 1902 until his death in 1909 ([187], p. 58).

C. S. Helrich, *The Quantum Theory—Origins and Ideas*, History of Physics,
https://doi.org/10.1007/978-3-030-79268-8_7

In June 1922 Niels Bohr gave the Wolfskehl lectures in Göttingen, later called the *Bohr Festspiele zu Göttingen* (Bohr-Festival at Göttingen). Sommerfeld invited Heisenberg to come with him to the lectures. At one point in one of the lectures, Bohr spoke of the work of Kramers on which Heisenberg had reported in the Sommerfeld seminar at LMU. Bohr said that the results were correct and would be shown to be so by experiment. At that point Heisenberg stood up and objected, saying that at Munich they had studied this work and found reason to doubt its validity. After the discussion that followed Bohr invited Heisenberg to accompany him on a walk on the Hainberg that overlooked Göttingen. This began a close relationship between Bohr and Heisenberg, who subsequently spent extended periods in Copenhagen ([186], pp. 128–132).

Matrices had only rarely been used by physicists prior to 1925 ([148], p. 217). For example, Born had collaborated with Theodore von Kármán (1881–1963) on studies of lattice vibrations in 1912 and they had employed a matrix formulation there. Another example was Minkowski's space-time, a geometrical version of Einstein's relativity, in which the Lorentz transformation was expressed as a four-dimensional matrix operator [197, 198]. However, Minkowski's matrix formulation had not completely convinced Sommerfeld. In his first paper on relativity, published in 1910, Sommerfeld wrote that he was making the transformations as geometric as possible and would provide a complete substitute for the matrix calculus introduced by Minkowski ([187], p. 36). On the other hand, Einstein fully adopted the matrix formulation in his paper on general relativity in 1916 [99].

The standard text on matrix theory available to German readers after 1910 was the German translation of Bôcher's book *Introduction to Higher Algebra* [13, 14]. Then in 1924 Richard Courant (1888–1972) published his two volume set *Methoden der mathematischen Physik*, based on the lectures by Hilbert, which appeared in English translation as *Methods of Mathematical Physics* in 1953. Among Courant's assistants in this task was Jordan, who was then completely familiar with matrices and column vectors, to which the first chapter in the set of books was devoted [60, 61].

After a description of the background that was instrumental in producing the matrix version of quantum mechanics, we will provide a detailed description of the three foundational papers that introduced the new theory. We have added to these the paper by Kramers and Heisenberg, which was important in forming Heisenberg's ideas, and an introduction to the work of the young English physicist Paul A.M. Dirac (1902–1984), whose methods have become central in quantum mechanics.

7.2 Background

Figure 7.1 is a modern (2006) photograph of the Great Hall (Aula) of the Georg-August University in Göttingen. The matrix version of quantum mechanics, which at the time was simply called quantum mechanics, was developed here.

Fig. 7.1 Great Hall (Aula) of the Georg-August University of Göttingen

In July of 1924 Heisenberg was appointed *Privatdozent* at Göttingen ([136], p. 87). He spent much of the year in Copenhagen with Bohr, and produced a paper on optical dispersion with Kramers, returning to Göttingen for the summer semester of 1925. In Copenhagen he had been influenced by some of Bohr's ideas, as well as by his work with Kramers. Now, back in Göttingen, Heisenberg began working on the intensities of hydrogen spectral lines, which was a topic that had not yet caught Born's attention. According to Heisenberg, he began by following the method he and Kramers had used in Copenhagen, but this ended in an impenetrable thicket of mathematical formulae and he could find no way through it ([136], p. 87).

Then in May Heisenberg's hayfever made it impossible for him to carry on. He asked Born for permission to spend two weeks on the island of Helgoland in the North Sea in order to get over the hayfever. Born granted the request and Heisenberg settled into an upstairs room in a guest house on Helgoland with a view over the sea. Sitting on the balcony, Heisenberg wrote, reminded him of what Bohr once said. To look at the sea gives us a certain sense of the infinite ([136], p. 88; [186], pp. 248, 249).

It took only a few days for Heisenberg to see his way through the mathematical complexities and to find a simpler approach to his problem. It then became clear to him that in the new physics only observable quantities should have any role to play. These needed to appear in the place of the Bohr–Sommerfeld quantum conditions. This additional requirement was the central point in the new theory that was to be formulated. In the way forward, there was no additional freedom ([136], p. 88).

It was not clear, however, that such a scheme would be devoid of contradiction. In particular, it was not obvious that this approach would be consistent with Bohr's statement (*Energiesatz*) that the energy difference between atomic levels was equal to $h\nu$, where ν was the frequency of the emitted radiation. Heisenberg could simply not escape the fact that, without this energy condition, the whole scheme would be worthless.[1] ([136]. p. 89).

Then one evening his method yielded one of the energy terms. He became excited, as he recalled, and began to make mathematical errors as he rapidly sought the other energy terms. Then at three o'clock in the morning the whole calculation lay before him. Energy conservation was valid in each circumstance. Sleep was out of the question. So he waited for the break of day to go out and climb a towering rock and watch the sunrise ([136], pp. 89–90).

Heisenberg admitted that what he had discovered was little more than a narrow rocky ledge in the mountains. But he sent his ideas to the highly critical Pauli, who encouraged him. And in Göttingen, Born and Jordan accepted his insight. In the words of Born, Heisenberg cut the Gordian knot by deciding that only measurable concepts should be used [28].

However, Heisenberg's revelation was not universally accepted. There is a letter from Heisenberg dated November 1925 in the Einstein Archive, replying to a note from Einstein (now lost). It is evident that the note contained many objections to Heisenberg's work. Then in April of 1926 Heisenberg gave a two hour lecture on quantum mechanics in von Laue's physics colloquium at the University of Berlin. After the lecture Einstein invited Heisenberg to his apartment for a discussion about the topic. The conversation as they walked to Einstein's apartment was casual and friendly. But once in his apartment Einstein began a pointed critique of Heisenberg's philosophical position.

Einstein pointed out that Heisenberg had accepted the existence of electrons, but not the trajectories. Einstein found this strange. Heisenberg attempted to clarify his position about our ability to observe. Einstein responded: "But you don't seriously believe that only actual magnitudes of observations can be included in a theory." Heisenberg expressed surprise here: given Einstein's definition of time in his relativity theory, he expected Einstein to be more accepting of his position. As Heisenberg recalled, Einstein admitted that he might then have used this philosophical position, but insisted that it was still nonsense. And then Einstein pointed out that it is theory that tells us what can be observed.[2] ([136], pp. 91–92; [144]).

Before leaving for Helgoland, Heisenberg had not discussed his work with Born to any great extent. He had some discussions with Born early in May when he was guessing at the intensities of the Balmer lines, and Heisenberg recalled that Born

---

[1] In Heisenberg's proposal, the Bohr energy frequency was

$$\nu\,(n, n-\alpha) + \nu\,(n-\alpha, n-\alpha-\beta) = \nu\,(n, n-\alpha-\beta)\,,$$

and the Bohr–Sommerfeld quantum requirement produced

$$h = 4\pi m \sum_{\alpha=0}^{\infty} \left\{ |a\,(n+\alpha, n)|^2\, \omega\,(n+\alpha, n)\, |a\,(n, n-\alpha)|^{\,2} \omega\,(n, n-\alpha) \right\},$$

with $\omega = 2\pi\nu$. The elements $a\,(n+\alpha, n)$ must result in consistent values for the frequencies

[2] This remark Heisenberg would later recall in his analysis of the trajectory of the electron in the cloud chamber, when he developed the principle of indeterminacy.

had given him a book on Bessel functions. After his stay on Helgoland, he had only communicated by letter with Pauli, and he sent the manuscript of the paper that emerged out of the effort on Helgoland first to Pauli in Hamburg, before showing it to anyone in Göttingen. He only went to see Born after receiving a favorable response from Pauli ([136], p. 90).

Heisenberg himself was still not certain that the paper should be published. He took the manuscript to Born and asked him to read it and to decide whether it should be published. At he same time, Born said, Heisenberg also asked for leave for the remainder of the academic term, which ended on 1 August. Heisenberg had been invited to lecture at the Cavendish Laboratory in Cambridge ([187], p. 7).

Various things, including a general weariness, kept Born from picking up the manuscript immediately. When he did, he later confessed, he was fascinated. Heisenberg had written the amplitudes of the line intensities as

$$C\left(n, n-\beta\right) = \sum_{\alpha=-\infty}^{+\infty} A\left(n, n-\alpha\right) B\left(n-\alpha, n-\beta\right), \tag{7.1}$$

where A and B were transition amplitudes. Born later wrote that he thought about this all day and into the night. Then in the morning he saw the light: this was matrix multiplication. He recalled his student days and a linear algebra course. What Heisenberg had formulated in (7.1) would normally have been written in the form

$$C_{nm} = \sum_{n'=-\infty}^{+\infty} A_{nn'} B_{n'm} \,. \tag{7.2}$$

In his paper Heisenberg had written the coordinate as

$$q\left(n, n'\right) = a\left(n, n'\right) \exp\left[i\omega\left(n, n'\right) t\right] \tag{7.3}$$

and the momentum, for the system he was considering, was simply

$$p\left(n, n'\right) = m\frac{\mathrm{d}q\left(n, n'\right)}{\mathrm{d}t} \,. \tag{7.4}$$

The quantum condition, which was based on the action requirement of the Bohr–Sommerfeld theory, viz.,

$$nh = \oint p\mathrm{d}q \,, \tag{7.5}$$

then took the form

$$h = 4\pi m \sum_{\alpha=0}^{\infty} \left\{ \left|a\left(n, n+\alpha\right)\right|^2 \omega\left(n+\alpha, n\right) - \left|a\left(n, n-\alpha\right)\right|^2 \omega\left(n, n-\alpha\right) \right\} . \tag{7.6}$$

Born reformulated this as

$$\frac{h}{2\pi i} = \sum_{n'=-\infty}^{\infty} (p_{nn'} q_{n'n} - q_{nn'} p_{n'n}) \ . \tag{7.7}$$

He then found himself looking at what he referred to as the "peculiar equation"

$$\mathbf{pq} - \mathbf{qp} = \frac{h}{2\pi i}\mathbf{1}, \tag{7.8}$$

where **q** and **p** were matrices representing the coordinate and momentum and **1** was the identity matrix ([187], p. 10, [28]).

Because of his rather extensive work with matrices, we may ask why it took any time at all for Born to recognize the fact that the square arrays Heisenberg had introduced were matrices. Jagdish Mehra and Helmut Rechenberg offer an explanation. They suggest this may lie in Born's scientific temperament. Whenever Born undertook a new study, he stuck with the standard approach to that study before turning to any new formulation. However, they also admit that the real reason may lie in two other issues. The first was that Heisenberg had discovered the noncommutative property of position and momentum in quantum mechanics and, second, that the noncommutative property of matrices had not featured in any of Born's previous applications of matrix algebra ([187], p. 43).

However, recognizing the connection at least allowed Born to more readily bring to bear his formidable mathematical prowess. On Sunday 19 July, Born took the train North to Hanover for the meeting of the Niedersachsen (Lower Saxony) Section of the DPG. Göttingen is in the southernmost tip of Sachsen and Hanover is almost directly north of it. Pauli, who had once been Born's assistant, was among those on the train, and Born was able to join him in a compartment. There Born raised the issue of Heisenberg's ideas and matrices, with which Pauli was already familiar, and asked if Pauli would be interested in joining with him in further development of those ideas. According to Born's recollection, Pauli responded coldly to the suggestion, implying that Born's love of tedious and complicated formalisms might spoil Heisenberg's physical ideas.

Max Jammer (1915–2010) mentions that Jordan was sitting in that same compartment and heard Born's enthusiasm for the progress Heisenberg had made. On the platform in Hanover, Jordan explained his experience in handling matrices with Courant, and expressed interest in becoming Born's assistant in the work ahead. Jordan had received his doctorate under Born in 1924 and was happy to work again with Born ([187], pp. 11–12).

The situation was awkward. Heisenberg had trusted Born with the decision to publish his work. However, he had not explicitly given Born permission to pursue an extension of the work. But Heisenberg was not present. After his time in Cambridge, he was planning a summer vacation in Munich and then a stay in Copenhagen. Any work that Born and Jordan did together must preserve Heisenberg's ideas and communication with Heisenberg would have to be by letter ([187], pp. 62–63).

Born was also planning a vacation in Silvaplana, Switzerland, with a stop at the Eberhard Karls University in Tübingen on the way. He and Jordan would only have a few days together before Born would leave on 30 July. He was also planning a stop at the ETH in Zurich on the return journey and would not be in Göttingen before the middle of September. Until then, the work would be carried out by Jordan.

At the end of July 1925 Born sent Heisenberg's paper to *Zeitschrift für Physik* and went to Switzerland for his vacation. Jordan remained in Göttingen and began work on the joint paper he would write with Born ([187], p. 59). Born informed Heisenberg of the new collaboration with Jordan in early August. And on 20 August, Heisenberg wrote to Jordan from Munich, saying that he no longer had the galley proofs that Born had said Jordan would like to see, but would send the manuscript ([187], p. 64).

Heisenberg also expressed great interest in the progress Jordan had made, particularly on the proof of the frequency condition. Jordan had proofs of energy conservation and the frequency condition for transitions. Heisenberg asked Jordan to send the proof to him in Copenhagen. If the proof turned out to be correct, Heisenberg wrote that he would be happy for Jordan to publish it.

Born was back in Göttingen soon after the middle of September, and the first paper by Born and Jordan, *Zur Quantenmechanik*, was received by *Zeitschrift für Physik* on 27 September ([187], pp. 62–64). The next step would be the *Drei-Männer-Arbeit* (three-man paper), as Born liked to call it, which would include Heisenberg ([148], p. 221).

In the following section, we discuss the original papers by Heisenberg, Born, and Jordan, and Born, Heisenberg, and Jordan, remaining as close to the original form as possible. In each paper we have dropped the final sections, which were devoted to applications. Specifically, we have not limited ourselves to simply outlining the papers. The paper by Kramers and Heisenberg, however, we have only discussed historically because it had a considerable influence on the ideas Heisenberg was forming.

Finally, we have introduced the thoughts of Dirac, based in part on his encounter with Heisenberg. We take the opportunity there to introduce the mathematical formalism Dirac developed as a doctoral student.

7.3 Matrix Mechanics

7.3.1 Kramers and Heisenberg

Heisenberg first settled in at Copenhagen in March of 1924. There he engaged with Kramers on the problem of optical dispersion. Kramers and Heisenberg published their work in January of 1925 [163]. What is important to us here is the fact that this work provided some of the background for Heisenberg's thinking that led to his first work on matrix mechanics [263].

Optical dispersion is a phenomenon that has long been understood in terms of the frequency of light and the index of refraction of the material through which the light passes. This results in the separation of light into different frequencies in rainbows and prisms. In the early 1870s, August Kundt (1839–1894) had discovered that, after passing through certain media, the light frequency was not continuous, but had gaps characteristic of the prismatic medium. In 1915, Debye, then Sommerfeld's assistant in Munich, attempted an explanation of these frequency gaps in hydrogen, using the Bohr model as starting point. However, Sommerfeld was satisfied with a classical interpretation of dispersion, with gaps related statistically to electron resonances.

Rudolf Ladenburg (1882–1952), a theoretically competent experimental physicist at the University of Breslau, knew that Sommerfeld's ideas were not correct. He had been studying dispersion in gases before 1913 and had identified a connection between the gaps in the dispersion frequency and the spectral lines of the medium. After WWI, Ladenburg became a staunch supporter of Bohr's atomic model. In 1921 he proposed a reinterpretation (*Umdeutung*) of dispersion based on two main points. The frequencies were identified with the frequencies of the Bohr model and the atomic tendency to undertake quantum jumps was identified with the Einstein probability coefficients [100]. This changed optical dispersion into a quantum phenomenon. The continuous spread of frequencies (colors) remained unexplained.

Ladenburg's ideas in 1921 were noticed at the newly established Institute for Theoretical Physics in Copenhagen, which Bohr directed. In 1923 Bohr pointed out that Ladenburg's paper indicated a connection between classical and quantum mechanics, in that the atoms were behaving like harmonic oscillators with characteristic frequencies matching the radiation. Bohr's assistant Kramers was the first to consider this idea based on Bohr's correspondence principle. He obtained a two-term dispersion formula with the terms representing absorption and emission of light. The papers Kramers had previously published on this idea were short and lacked mathematical development.

During his stay in Copenhagen, Heisenberg worked with Kramers in an attempt to consolidate the mathematics of optical dispersion using Bohr's model. Heisenberg was very interested to be involved in the basic atomic work going on at Copenhagen, and was therefore happy to be involved with Kramers.

We know from the extensive analysis of Mehra and Rechenberg that the relationship between Kramers and Heisenberg was not always easy. Each brought expectations to the partnership, some of which required Bohr's assistance in their resolution. This work was also taking place before the demise of the BKS paper at the hands of Bothe and Geiger (see Sect. 5.4). Kramers was himself unconvinced of that demise, even after the experimental results in the spring of 1925. The reception of the work by Kramers and Heisenberg was positive, however.

In this collaboration, Heisenberg had seen the close connection between the Fourier components in classical physics and the quantum theoretical quantities. Moreover, he came to see that it was the Fourier components that really mattered, not the atomic orbits. The problem was just to establish the connection between those Fourier components of the classical theory and the ideas of the action and angle variables that were emerging from what we may term the old quantum theory ([186], pp. 170–190; [263]).

7.3.2 Heisenberg

The title of this paper is *Über quantentheoretische Umdeutung kinematischer und mechanischer Beziehungen* (Quantum theoretical re-interpretation of kinematic and mechanical relations). What we present here is taken directly from the original paper.

The abstract for the paper, which Born submitted to *Zeitschrift der Physik* in July 1925, is a single sentence [135]:

> This work is an attempt to obtain the principles of a quantum theoretical mechanics based solely on the relationship among observable quantities.

Heisenberg then began by pointing out that in the quantum theory the formal rules for calculating the observable quantities, such as energies, are based on other quantities, such as electron trajectories, which are not observable. These rules are in jeopardy unless we hold fast to the hope that the unobservable quantities will eventually be observable. In reality, however, the complex phenomena associated with the interaction of electrons with electromagnetic waveforms dispel these hopes.

There was a tendency to think about quantum phenomena as deviations from the basic classical theory. Considerations of basic principles, such as the Bohr–Einstein frequency relationship,[3] which completely defies any classical understanding, quickly put an end to that belief.

Faced with this situation Heisenberg found it advisable to construct a quantum theoretical mechanics which was analogous to classical mechanics but was based solely on observable quantities. In this paper, Heisenberg presented some new quantum mechanical relationships and attempted a complete treatment of a few special problems. He limited his treatment to a single dimension.

After a few words on the complications encountered in the interactions of electrons with electromagnetic waves as described by Maxwell's theory, he asked what we should expect the higher-level terms in the quantum theory to look like. Since classically we simply calculate the terms to some approximation, while representing the motion of the electron in terms of a Fourier series or a Fourier transformation, we should expect something similar in the quantum theory. However, this question of representation has nothing to do with the dynamics of the electron. This translation from classical to quantum theory seemed to Heisenberg to be a straightforward kinematic question. It was a question of representation that would be answered by finding quantum mechanical terms that would correspond to the Fourier terms in the representation of $x(t)$. Then he asked rhetorically what could be done with $x^2(t)$.

[3] This is the claim that a transition between the energy levels $\mathcal{E}_n$ and $\mathcal{E}_m$ produces light with frequency $\nu = (\mathcal{E}_n - \mathcal{E}_m)/h$.

Heisenberg began his discussion by picking out some classical and quantum terms he would need to consider. These began with the Bohr–Einstein relation, which he wrote as

$$\nu(n, n-\alpha) = \frac{1}{h}\{W(n) - W(n-\alpha)\}, \tag{7.9}$$

where $W(n)$ and $W(n-\alpha)$ are the energies of the quantum states n and $n-\alpha$. He compared this with the relation between the *action* and *angle variables* of classical mechanics. The frequencies of the motion of a classical system are

$$\nu(n, \alpha) = \alpha\nu(n) = \alpha\frac{\partial\mathcal{H}(J)}{\partial J_{\mathrm{n}}}, \tag{7.10}$$

where J_{n} are the angle variables and the frequencies $\nu(n, \alpha)$ are the time rates of change of the angle variables ([140], pp. 190–192).

He also noted that classically (in a Fourier series) the frequencies are related as

$$\nu(n, \alpha) + \nu(n, \beta) = \nu(n, \alpha + \beta), \tag{7.11}$$

while quantum mechanically,

$$\nu(n, n-\alpha) + \nu(n-\alpha, n-\alpha-\beta) = \nu(n, n-\alpha-\beta) \tag{7.12}$$

and

$$\nu(n-\beta, n-\alpha-\beta) + \nu(n, n-\beta) = \nu(n, n-\alpha-\beta), \tag{7.13}$$

which follow from energy conservation and (7.9).

In addition to the question of the quantum theoretical representation of the frequencies, there was also the representation of the amplitudes of the terms in the Fourier series (or integral). Classically, the amplitudes result from the approximations used in the calculation. They may be rather complex quantities depending on variables such as the polarization and phase, in addition to the indices n and α. He then noted that he would use the quantum theoretical amplitude

$$\mathrm{Re}\{a(n, n-\alpha)\exp[i\omega(n, n-\alpha)t]\} \tag{7.14}$$

in place of the classical amplitude[4]

$$\mathrm{Re}\{a_\alpha(n)\exp[i\alpha\omega(n)t]\}. \tag{7.15}$$

For the one-dimensional situation he was considering, Heisenberg had a representation for the classical motion as a Fourier series

[4] Here, Heisenberg identified the amplitudes as real parts of the coefficients.

$$x(n,t) = \sum_{\alpha} a_{\alpha}(n) \exp[i\alpha\omega(n)t] \tag{7.16}$$

or as a Fourier transform

$$x(n,t) = \int_{-\infty}^{+\infty} a_{\alpha}(n) \exp[i\alpha\omega(n)t]\, d\alpha. \tag{7.17}$$

There was a logical difficulty in writing down an expression for a quantum mechanical representation of $x(n,t)$ because of the equal weights of n and $n-\alpha$. Nevertheless, he could still use the totality of the terms $a\,(n, n-\alpha) \exp[i\omega\,(n, n-\alpha)\,t]$ for the quantum representation of the quantity $a_{\alpha}(n) \exp[i\alpha\omega(n)t]$. That is, in going from classical to quantum theory, he changed the Fourier term to

$$a_{\alpha}(n) \exp[i\alpha\omega(n)t] \rightarrow a\,(n, n-\alpha) \exp[i\omega\,(n, n-\alpha)\,t]\,. \tag{7.18}$$

Keeping to measurable quantities, the state of motion $x\,(n,t)$ was describable in terms of transitions to other stationary states. The correspondence principle then required that the α th component of the series (7.16) should correspond to a jump from state n to state $n-\alpha$ with spectral frequency $\omega\,(n, n-\alpha)$ [106, 107].

Once again, he asked how $x^2(t)$ should be represented. But now he answered the question. Classically, one would like to write

$$x^2(n,t) = \sum_{\beta} b_{\beta}(n) \exp[i\beta\omega(n)t]\,, \tag{7.19}$$

which comes from the classical product of Fourier series

$$x^2(n,t) = \sum_{\alpha}\sum_{\beta} a_{\alpha}\,(n)\, a_{\beta}(n) \exp[i\,(\alpha+\beta)\,\omega(n)t]\,. \tag{7.20}$$

In quantum theoretical terms, the simplest and most natural assumption would be to set

$$\begin{aligned} &b\,(n, n-\beta) \exp[i\omega\,(n, n-\beta)\,t] \\ &= \sum_{\alpha} a\,(n, n-\alpha)\, a\,(n-\alpha, n-\beta) \exp[i\omega\,(n, n-\beta)\,t]\,. \end{aligned} \tag{7.21}$$

Heisenberg followed this with a formulation for $x^3(t)$. There the coefficient $c\,(n,\gamma)$, which was classically

$$c\,(n,\gamma) = \sum_{\alpha}\sum_{\beta} a_{\alpha}\,(n)\, a_{\beta}(n) a_{\gamma-\alpha-\beta}(n)\,,$$

became $c\,(n, n-\gamma)$ quantum theoretically, where

$$c\,(n, n-\gamma) = \sum_{\alpha}\sum_{\beta} a\,(n, n-\alpha)\, a\,(n-\alpha, n-\alpha-\beta)\, a\,(n-\alpha-\beta, n-\gamma)\,.$$

This set the pattern for products of $x(t)y(t)$.

Heisenberg then undertook a more detailed study of the one-dimensional motion resulting from a general force for which the Newtonian equation of motion was

$$\ddot{x} + f\,(x) = 0. \tag{7.22}$$

The original position $x(n,t)$ and the frequency $\omega\,(n)$ were classical terms provided by the solution to the classical equations of motion. Heisenberg then introduced the quantum condition of the old quantum theory (see (4.19) in Sect. 4.3) in the form

$$J = \oint p\mathrm{d}x = nh. \tag{7.23}$$

This is the action variable of classical mechanics. An action variable is associated with each canonical coordinate of the mechanical system. In spite of the terminology, the action variables of a mechanical system depend only on the constants of the motion and are, therefore, themselves constants of the motion.

The variables conjugate to the action variables are the angle variables, which carry the relationship between the coordinates and the action variables. Action and angle variables are important when studying periodic motion in complex systems. In general, the energy of a conservative system may be expressed as a function of the action variables $\mathcal{E} = \mathcal{E}\,(J)$ ([140], Sect. 5.7).

When represented as a Fourier series, the classical solution of (7.22) is

$$x(t) = \sum_{\alpha} a_{\alpha}(n) \exp\left[i\alpha\omega_{\mathrm{n}}t\right], \tag{7.24}$$

and the time derivative of (7.16) is

$$\dot{x}(t) = \sum_{\alpha} a_{\alpha}(n)\,(i\alpha\omega_{\mathrm{n}}) \exp\left[i\alpha\omega_{\mathrm{n}}t\right]. \tag{7.25}$$

The system Heisenberg was considering had only a single dimension, and hence a single action. With the solution in the Fourier representation (7.25), Eq. (7.23) may be integrated over a period of the motion to obtain

$$J = nh = \oint p\mathrm{d}x = \oint m\dot{x}^2\mathrm{d}t = 2\pi m \sum_{\alpha} \left|a_{\alpha}(n)\right|^2 \alpha^2\omega_{\mathrm{n}}. \tag{7.26}$$

Then

$$h = \frac{\mathrm{d}}{\mathrm{d}n}(nh) = 2\pi m \sum_{\alpha} \alpha \frac{\mathrm{d}}{\mathrm{d}n}\left(\alpha\omega_{\mathrm{n}} \left|a_{\alpha}(n)\right|^2\right),$$

which is

$$h = 4\pi m \sum_{\alpha=0}^{\infty} \left\{ |a(n+\alpha, n)|^2 \, \omega(n+\alpha, n) - |a(n, n-\alpha)|^2 \omega(n, n-\alpha) \right\}. \tag{7.27}$$

We may now recognize this as (7.6), which Born manipulated into the peculiar equation (7.8) he eventually found himself looking at

In the last section of the paper, Heisenberg considered the anharmonic oscillator. [135] We shall not consider this final section here.

7.3.3 *Born and Jordan*

This paper is entitled *Zur Quantenmechanik* (On Quantum Mechanics) and may be thought of as *Zur Quantenmechnik I*, since the subsequent publication by Born, Heisenberg, and Jordan had the title *Zur Quantenmechanik II*. What we present here is taken directly from the original paper *Zur Quantenmechanik* [23].

The first sentences of the abstract of the paper written by Born and Jordan (B&J) are:

> The postulates recently proposed by Heisenberg will be developed into a systematic theory of quantum mechanics (initially for systems of one degree of freedom). The mathematical tool is the matrix calculus. Following a brief presentation of this calculus, the mechanical equations of motion will be derived from a variational principle, and it will be proved that the law of energy and Bohr's frequency condition follow from the mechanical equations on the grounds of Heisenberg's quantum condition.

The first section of the paper developed matrix analysis. The authors noted that they would "calculate with infinite square matrices" and cite Bôcher [14] and Courant and Hilbert [60]. Their treatment included time derivatives of matrices and partial derivatives of matrix products and matrix functions.

Time derivatives of matrices presented no problem. A matrix expression dependent on time $\mathbf{A}(t)$ had the time derivative

$$\frac{\mathrm{d}\mathbf{A}(t)}{\mathrm{d}t} = \dot{\mathbf{A}} = \lim_{\tau \to 0} \frac{\mathbf{A}(t+\tau) - \mathbf{A}(t)}{\tau}. \tag{7.28}$$

The time derivative of a product was

$$\frac{\mathrm{d}\mathbf{AB}}{\mathrm{d}t} = \dot{\mathbf{A}}\mathbf{B} + \mathbf{A}\dot{\mathbf{B}}, \tag{7.29}$$

and it followed that

$$\frac{d}{dt}(\mathbf{X}_1\mathbf{X}_2\cdots\mathbf{X}_n) = \dot{\mathbf{X}}_1\mathbf{X}_2\cdots\mathbf{X}_n + \mathbf{X}_1\dot{\mathbf{X}}_2\cdots\mathbf{X}_n + \cdots + \mathbf{X}_1\mathbf{X}_2\cdots\cdots\dot{\mathbf{X}}_n\,. \tag{7.30}$$

However, partial differentiation of matrix functions with respect to a matrix is not such a simple task. B&J were able to show that partial derivatives of matrix products such as $(\mathbf{Q}^n\mathbf{P}^m)$ could be given by

$$\frac{\partial}{\partial \mathbf{Q}}(\mathbf{Q}^n\mathbf{P}^m) = \mathbf{Q}^{n-1}\mathbf{P}^m + \mathbf{Q}^{n-2}\mathbf{P}^m\mathbf{Q} + \cdots + \mathbf{P}^m\mathbf{Q}^{n-1}\,. \tag{7.31}$$

The more general problem, however, was the partial derivative of a general matrix function of other matrices, such as $\mathbf{F}(\mathbf{Q}, \mathbf{P}, \ldots, \mathbf{R})$. Fortunately B&J were able to obtain the element of the partial derivative $(\partial\mathbf{F}/\partial\mathbf{Q})_{mn}$ of such a matrix function in terms of the trace of the matrix $\mathbf{F}(\mathbf{Q}, \mathbf{P}, \ldots, \mathbf{R})$. The trace is the sum of the diagonal terms in the matrix, which B&J denoted by $\mathbf{D}(\mathbf{F})$

$$\mathbf{D}(\mathbf{F}) = \sum_{n=0}^{\infty} f_{nn}. \tag{7.32}$$

They proved that the element of the partial derivative $(\partial\mathbf{F}/\partial\mathbf{Q})_{mn}$ is

$$\left(\frac{\partial\mathbf{F}}{\partial\mathbf{Q}}\right)_{mn} = \frac{\partial\mathbf{D}(\mathbf{F})}{\partial q_{nm}}. \tag{7.33}$$

The second section of the paper was devoted to dynamics. B&J accepted Heisenberg's coordinates, which they denoted by $\mathbf{q}$, and with which they associated the conjugate momenta $\mathbf{p}$. These were the matrices

$$\mathbf{q} = (q\,(nm)\exp[2\pi i\nu\,(nm)\,t]) \tag{7.34}$$

and

$$\mathbf{p} = (p\,(nm)\exp[2\pi i\nu\,(nm)\,t])\,, \tag{7.35}$$

for which they used a round bracket notation rather than the Heisenberg's curly bracket notation. The frequencies $\nu\,(nm)$ were the quantum-theoretical frequencies for the transitions between the quantum states n and m. These matrices were to be *Hermitian* (for all time t) and infinite in dimension.[5] Denoting the complex conjugate by an asterisk, the Hermitian requirement is

[5] A Hermitian matrix is equal to its *adjoint*, which is the complex conjugate of its *transpose*. The transpose is formed by interchanging rows and columns in the matrix. That is, for a Hermitian matrix $\mathbf{R}$, the elements $R_{\mu\nu}$ satisfy

$$R_{\mu\nu} = R^*_{\nu\mu}.$$

$$q\,(nm)\,q\,(nm)^* = q\,(nm)\,q\,(mn) = |q\,(nm)|^2 \tag{7.36}$$

and

$$\nu\,(nm) = -\nu\,(mn) \quad \text{with} \quad \nu\,(nn) = 0. \tag{7.37}$$

If q is Cartesian, (7.36) will be the probability of the transition $n \leftrightarrows m$.

B&J also required the frequencies to satisfy Heisenberg's summation axiom

$$\nu\,(jk) + \nu\,(kl) + \nu\,(lj) = 0. \tag{7.38}$$

They then required terms W_{n} such that

$$h\nu\,(nm) = W_{\mathrm{n}} - W_{\mathrm{m}}, \tag{7.39}$$

as did Heisenberg. Inherent in (7.39) is the assumption of a diagonal matrix $\mathbf{W}$ ($\mathbf{W} = W_{\mathrm{n}}\delta_{\mathrm{nm}}$). The validity of this assumption is in turn based on the possibility of diagonalizing the Hamiltonian matrix for the system, since the elements W_{n} are energies. Diagonalization of a matrix should not be considered a trivial task. John von Neumann (1903–1957) devoted considerable space to this problem in ([206, Chap. 1]), but B&J did not go into the details in the paper we are discussing here.

Using the rules of matrix addition and multiplication as well as (7.37)–(7.39), B&J observed that any function $\mathbf{g}\,(\mathbf{p}, \mathbf{q})$ takes the form

$$\mathbf{g} = (g\,(nm)\exp\left[2\pi i\nu\,(nm)\,t\right]). \tag{7.40}$$

From this it follows that the time-dependent factor $\exp\left[2\pi i\nu\,(nm)\,t\right]$ will be common to all terms. This factor may then be dropped and $\mathbf{g}$ may be written simply as

$$\mathbf{g} = (g\,(nm)). \tag{7.41}$$

For the time derivative,

$$\frac{\mathrm{d}\mathbf{g}}{\mathrm{d}t} = \dot{\mathbf{g}} = 2\pi i\,(\nu\,(nm)\,g\,(nm)). \tag{7.42}$$

If $\nu\,(nm) \neq 0$, as B&J preferred to assume, then $\dot{\mathbf{g}} = \mathbf{0}$ implies that the off-diagonal terms in $\mathbf{g}$ are zero, i.e., $\mathbf{g}$ is diagonal. Then $g\,(nm) = \delta_{\mathrm{nm}}g\,(nm)$.

B&J then assumed that the matrix $\mathbf{W}$ was diagonal, i.e., $\mathbf{W} = \delta_{\mathrm{nm}}W_{\mathrm{m}}$. Hence,

$$\begin{aligned}\mathbf{Wg} &= \sum_{\mathrm{k}} \delta_{\mathrm{nk}}W_{\mathrm{k}}g\,(km) = W_{\mathrm{n}}g\,(nm)\,,\\ \mathbf{gW} &= \sum_{\mathrm{k}} g\,(nk)\,\delta_{\mathrm{km}}W_{\mathrm{m}} = g\,(nm)\,W_{\mathrm{m}}.\end{aligned} \tag{7.43}$$

From (7.42) and (7.39), the time derivative of $\mathbf{g}$ is then

$$\dot{\mathbf{g}} = \frac{2\pi i}{h}\left(\left(W_{\mathrm{n}} - W_{\mathrm{m}}\right) g\,(nm)\right) = \frac{2\pi i}{h}\left(\mathbf{Wg} - \mathbf{gW}\right). \tag{7.44}$$

B&J showed that a simple permutation operation on all matrices does not change the form of this equation. However, general transformations may produce a non-diagonal **W** matrix. They noted that it would be profitable to study these general transformations because this could yield insights into the theory, and they promised to consider this in a later publication.

For the case of a Hamiltonian function of the form

$$\mathbf{H}\,(\mathbf{q}, \mathbf{p}) = \frac{1}{2m}\mathbf{p}^2 + \mathbf{U}\,(\mathbf{q})\ , \tag{7.45}$$

they assumed the canonical equations to be

$$\frac{\mathrm{d}\mathbf{q}}{\mathrm{d}t} = \dot{\mathbf{q}} = \frac{\partial \mathbf{H}}{\partial \mathbf{p}} \tag{7.46}$$

and

$$\frac{\mathrm{d}\mathbf{p}}{\mathrm{d}t} = \dot{\mathbf{p}} = -\frac{\partial \mathbf{H}}{\partial \mathbf{q}} = -\frac{\partial \mathbf{U}}{\partial \mathbf{q}}. \tag{7.47}$$

(see also [187], pp. 74–76; [148], p. 219) Classically, these canonical equations arise from an extremum of Hamilton's principal function

$$S = \int_{\mathrm{t_0}}^{\mathrm{t}} L\,(q, \dot{q}, t)\,\mathrm{d}t\ , \tag{7.48}$$

where the Lagrangian is defined as the difference in kinetic and potential energies for the system, viz.,

$$L\,(q, \dot{q}, t) = T\,(\dot{q}) - U(q). \tag{7.49}$$

The classical Hamiltonian

$$\mathcal{H}\,(q, p, t) = p\dot{q} - L\,(q, \dot{q}, t) \tag{7.50}$$

results from a Legendre transformation, replacing the velocity dependence by a dependence on the canonical momenta $p = \partial L/\partial \dot{q}$. The principal function is then

$$S = \int_{\mathrm{t_0}}^{\mathrm{t}} \left[p\dot{q} - \mathcal{H}\,(q, p, t)\right]\mathrm{d}t\ . \tag{7.51}$$

B&J suggested that if one carried out a Fourier series representation of the Lagrangian (written here in terms of the Hamiltonian) and considered a long time interval $t_1 - t_0$, only the constant terms in the Lagrangian would contribute to the integral. The principal function would thus take the form of the trace of the Lagrangian matrix, which is

$$\mathbf{D}(\mathbf{L}) = \mathbf{D}(\mathbf{p}\dot{\mathbf{q}} - \mathbf{H}(\mathbf{q}, \mathbf{p})) . \tag{7.52}$$

The canonical equations of motion would then be obtained by setting the derivatives of $\mathbf{D}(\mathbf{L})$ with respect to $(\mathbf{q}, \mathbf{p})$ equal to zero, while holding $\nu(nm)$ constant. Using (7.33), they found[6]

$$2\pi i \nu(nm) q(nm) = \frac{\partial \mathbf{D}(\mathbf{L})}{\partial p(mn)}$$

and

$$2\pi i \nu(nm) p(nm) = \frac{\partial \mathbf{D}(\mathbf{L})}{\partial q(mn)}$$

to be the equations for the matrix elements. In matrix form, these are (7.46) and (7.47), which they had assumed as the form of Hamilton's canonical equations.

B&J then turned to what they called the "classical" quantum theory:

$$J = \oint p\mathrm{d}q = \int_0^{1/\nu} p\dot{q}\mathrm{d}t . \tag{7.53}$$

Here, they noted that, upon integration, the classical Fourier series

$$\begin{aligned} p &= \sum_{\mathrm{k}} p_{\mathrm{k}} \exp(2\pi i k \nu t) , \\ q &= \sum_{\mathrm{k}} q_{\mathrm{k}} \exp(2\pi i k \nu t) , \end{aligned} \tag{7.54}$$

yield

$$1 = 2\pi i \sum_{\mathrm{k}} k \frac{\partial}{\partial J}(q_{\mathrm{k}} p_{\text{-k}}) \tag{7.55}$$

as the classical result.

In the matrix theory B&J were developing, they pointed out that there should be a correspondence between

$$\sum_{\mathrm{k}} k \frac{\partial}{\partial J}(q_{\mathrm{k}} p_{\text{-k}}) \text{ and } \frac{1}{h} \sum_{\tau} (q(n+\tau, n) p(n, n+\tau) - q(n, n-\tau) p(n-\tau, n)) , \tag{7.56}$$

in which all $q(nm)$ and $p(nm)$ with negative indices are set equal to zero. In this way they found that the quantum analog of (7.55) is

---

[6] The variation considered is one in which the mechanical states at the temporal endpoints t_0 and t_1 are fixed. This is referred to as a Lagrange or δ-variation. The frequencies are experimental data and therefore considered constant.

$$\sum_{\mathrm{k}} \left(p\left(nk\right)q\left(kn\right) - q\left(nk\right)p\left(kn\right)\right) = \frac{h}{2\pi i}. \tag{7.57}$$

Then B&J noted that, if they used $p = m\dot{q}$ in (7.57), they obtained

$$\sum_{\mathrm{k}} \nu\left(kn\right)\left|q\left(nk\right)\right|^2 = \frac{h}{8\pi^2 m}, \tag{7.58}$$

which coincides with the Heisenberg form (7.27) for the amplitudes.

In the penultimate section of the paper, entitled *Consequences: Laws of Frequency and Energy*, B&J noted that, using the above developments, they were able to provide a complete account of all the basic laws of the new mechanics, and in particular, the laws of conservation of energy and the Bohr frequency condition. Conservation of energy resulted from $\dot{\mathbf{H}} = 0$, or the fact that $\mathbf{H}$ is diagonal, which according to Heisenberg, meant that the diagonal terms were the energies of the various states in the system. The Bohr frequency condition, on the other hand, would require

$$h\nu\left(nm\right) = H\left(nn\right) - H\left(mm\right). \tag{7.59}$$

They began the discussion by defining

$$\mathbf{d} = \mathbf{pq} - \mathbf{qp}.$$

Then,

$$\dot{\mathbf{d}} = \dot{\mathbf{p}}\mathbf{q} + \mathbf{p}\dot{\mathbf{q}} - \dot{\mathbf{q}}\mathbf{p} - \mathbf{q}\dot{\mathbf{p}} = \mathbf{q}\left(\frac{\partial \mathbf{H}}{\partial \mathbf{q}} - \frac{\partial \mathbf{H}}{\partial \mathbf{q}}\right)\mathbf{q} + \mathbf{p}\left(\frac{\partial \mathbf{H}}{\partial \mathbf{p}} - \frac{\partial \mathbf{H}}{\partial \mathbf{p}}\right)\mathbf{p}.$$

Therefore, $\dot{\mathbf{d}} = 0$. From (7.42), the time derivative of $\mathbf{d}$ is

$$\dot{\mathbf{d}} = \sum_{\mathrm{k=0}}^{\infty} 2\pi i\nu\left(nm\right)\left[d\left(nm\right)\right] = \mathbf{0}. \tag{7.60}$$

Since if $n \neq m$, the frequency $\nu\left(nm\right) \neq 0$, it follows that $d\left(nm\right) = 0$. If n and m are equal, the frequency $\nu\left(nn\right) = 0$ and $d\left(nn\right)$ is arbitrary. The matrix $\mathbf{d}$ is then diagonal and (7.57) becomes

$$\mathbf{pq} - \mathbf{qp} = \frac{h}{2\pi i}\mathbf{1}. \tag{7.61}$$

The proof was based on the canonical equations, so the quantum equations of motion required this noncommutation result. This was what B&J now called the *exact (sharpened) quantum condition*. This is the point at which Planck's constant h entered the theory. B&J based all further conclusions on (7.61).

They began with the equations

$$\mathbf{p}^{n}\mathbf{q} - \mathbf{q}\mathbf{p}^{n} = n\frac{h}{2\pi i}\mathbf{p}^{n-1} \tag{7.62}$$

and

$$\mathbf{p}\mathbf{q}^{n} - \mathbf{q}^{n}\mathbf{p} = n\frac{h}{2\pi i}\mathbf{q}^{n-1}, \tag{7.63}$$

which are readily proven from (7.61) by induction. From here they proceeded to prove the conservation of energy and the Bohr frequency condition.

They first considered a separable Hamiltonian

$$\mathbf{H}(\mathbf{q}, \mathbf{p}) = \mathbf{H}_1(\mathbf{p}) + \mathbf{H}_2(\mathbf{q}), \tag{7.64}$$

in which both $\mathbf{H}_1$ and $\mathbf{H}_2$ are represented as series in their arguments $\mathbf{p}$ and $\mathbf{q}$, i.e.,

$$\mathbf{H}_1(\mathbf{p}) = \sum_{n} a_n \mathbf{p}^{n} \text{ and } \mathbf{H}_2(\mathbf{q}) = \sum_{n} b_n \mathbf{q}^{n}. \tag{7.65}$$

Hence,

$$\begin{aligned}
\mathbf{Hq} - \mathbf{qH} &= \sum_{n} \left(a_n \mathbf{p}^{n}\mathbf{q} + b_n \mathbf{q}^{n+1}\right) - \sum_{n} \left(a_n \mathbf{q}\mathbf{p}^{n} + b_n \mathbf{q}^{n+1}\right) \\
&= \sum_{n} a_n \left(\mathbf{p}^{n}\mathbf{q} - \mathbf{q}\mathbf{p}^{n}\right) \\
&= \frac{h}{2\pi i}\sum_{n} n a_n \mathbf{p}^{n-1} \\
&= \frac{h}{2\pi i}\frac{\partial \mathbf{H}}{\partial \mathbf{p}},
\end{aligned} \tag{7.66}$$

and, in the same fashion,

$$\mathbf{Hp} - \mathbf{pH} = -\frac{h}{2\pi i}\frac{\partial \mathbf{H}}{\partial \mathbf{q}}. \tag{7.67}$$

Using the canonical equations (7.46) and (7.47), this led to

$$\begin{aligned}
\dot{\mathbf{q}} &= \frac{2\pi i}{h}(\mathbf{Hq} - \mathbf{qH}) \\
\dot{\mathbf{p}} &= \frac{2\pi i}{h}(\mathbf{Hp} - \mathbf{pH}).
\end{aligned} \tag{7.68}$$

For a general matrix function $\mathbf{f}(\mathbf{q}, \mathbf{p})$, which can be written as a power series in $\mathbf{q}$ and $\mathbf{p}$, i.e.,

$$\mathbf{f}(\mathbf{q}, \mathbf{p}) = \sum_{n} \left(\alpha_n \mathbf{q}^{n} + \beta_n \mathbf{p}^{n}\right), \tag{7.69}$$

B&J showed that

$$\mathbf{fq} - \mathbf{qf} = \frac{h}{2\pi i}\frac{\partial \mathbf{f}}{\partial \mathbf{p}} \tag{7.70}$$

and

$$\mathbf{pf} - \mathbf{fp} = \frac{h}{2\pi i}\frac{\partial \mathbf{f}}{\partial \mathbf{q}} . \tag{7.71}$$

To prove (7.70), we first note that

$$\frac{\partial \mathbf{f}}{\partial \mathbf{p}} = \sum_{\mathrm{n}} b_{\mathrm{n}} n \mathbf{p}^{\mathrm{n\text{-}1}}$$

and

$$\begin{aligned}\mathbf{fq} - \mathbf{qf} &= \sum_{\mathrm{n}} \left(a_{\mathrm{n}}\mathbf{q}^{\mathrm{n+1}} + b_{\mathrm{n}}\mathbf{p}^{\mathrm{n}}\mathbf{q}\right) - \sum_{\mathrm{n}} \left(a_{\mathrm{n}}\mathbf{q}^{\mathrm{n+1}} + b_{\mathrm{n}}\mathbf{q}\mathbf{p}^{\mathrm{n}}\right) \\ &= \sum_{\mathrm{n}} \left(b_{\mathrm{n}}\mathbf{p}^{\mathrm{n}}\mathbf{q} - b_{\mathrm{n}}\mathbf{q}\mathbf{p}^{\mathrm{n}}\right) .\end{aligned}$$

Equation (7.70) follows if

$$\mathbf{p}^{\mathrm{n}}\mathbf{q} - \mathbf{q}\mathbf{p}^{\mathrm{n}} = n\frac{h}{2\pi i}\mathbf{p}^{\mathrm{n\text{-}1}},$$

which is (7.62) and was proved above. The proof of (7.71) is identical. Then,

$$\dot{\mathbf{f}} = \frac{2\pi i}{h}\left(\mathbf{Hf} - \mathbf{fH}\right) , \tag{7.72}$$

for the general separable vector function **f**.

From the canonical equations in the form (7.68), the fact that the Hamiltonian is diagonal implies

$$\begin{aligned} h\nu\,(nm)\, q\,(nm) &= \left(H\,(nn) - H\,(mm)\right) q\,(nm) \ , \\ h\nu\,(nm)\, p\,(nm) &= \left(H\,(nn) - H\,(mm)\right) p\,(nm) \ , \end{aligned} \tag{7.73}$$

which is the Bohr frequency condition.

The remainder of the chapter deals with more general forms of the Hamiltonian but we shall not consider these here.

The final section of the paper is devoted to the anharmonic oscillator, which Heisenberg also considered. The paper concluded with a discussion of the quantization of the electromagnetic field ([148], p. 220).

7.3.4 Born, Heisenberg, and Jordan

Here we discuss the paper entitled *Zur Quantenmechanik II* [24]. In the abstract Born, Heisenberg, and Jordan (BHJ) write:

> The quantum mechanics developed in Part I of this paper from Heisenberg's approach is extended here to systems having arbitrarily many degrees of freedom. [...] The results so obtained are employed in the derivation of [...] angular momentum conservation laws and selection rules [...] Finally, the theory is applied to the statistics of eigenvibrations of a blackbody cavity.

Since Born was determined to link the development to classical mechanics, it was logical to look at the path followed by Jacobi in his papers on Hamilton's approach. Central to Jacobi's thinking was the fact that a mechanical system must follow a trajectory in phase space defined by the requirement that Hamilton's canonical equations were valid at each point. Such a transformation is called a *canonical transformation*. In quantum mechanics the canonical equations (7.68) are a result of the exact quantum condition (7.61). Therefore, transformations that preserve the exact quantum condition will be canonical and the canonical equations will always hold.

After introducing the canonical transformation, BHJ were able to derive a perturbation theory that was similar to classical theory. Here, they sought a simplicity comparable to classical theory, but which depended only on observables. It was their contention that this was natural. At the end of the introduction they admitted, however, that the theory did not yet furnish solutions to the main difficulties of quantum theory.

In the first section, BHJ outlined the theory for systems having a single degree of freedom, beginning with

$$\mathbf{pq} - \mathbf{qp} = \frac{h}{2\pi i}\mathbf{1} \tag{7.74}$$

as a postulate. They first called this the *fundamental quantum mechanical relation*, and later the *commutation relation*. They defined matrix elements and dropped the time-dependent exponential as in Part I, defined the partial derivative with respect to a matrix component, and introduced the element of the partial derivative as

$$\frac{\partial \mathbf{f}}{\partial \mathbf{x}_1}(nm) = \frac{\partial \mathbf{D}(\mathbf{f})}{\partial x_1(mn)}. \tag{7.75}$$

They also introduced the equations

$$\mathbf{fq} - \mathbf{qf} = \frac{h}{2\pi i}\frac{\partial \mathbf{f}}{\partial \mathbf{p}} \tag{7.76}$$

and

$$\mathbf{pf} - \mathbf{fp} = \frac{h}{2\pi i}\frac{\partial \mathbf{f}}{\partial \mathbf{q}}, \tag{7.77}$$

which they had obtained in Part I.

In the next section, BHJ quoted the canonical equations

$$\dot{\mathbf{q}} = \frac{2\pi i}{h} (\mathbf{Hq} - \mathbf{qH}) \, ,$$
$$\dot{\mathbf{p}} = \frac{2\pi i}{h} (\mathbf{Hp} - \mathbf{pH}) \, . \tag{7.78}$$

Then for the frequencies, they introduced the summation relation

$$\nu\,(jk) + \nu\,(kl) + \nu\,(lj) = 0 \tag{7.79}$$

and the Bohr frequency condition, which they wrote in the form

$$\nu\,(nm) = \frac{W_{\mathrm{n}} - W_{\mathrm{m}}}{h} . \tag{7.80}$$

Then, for an arbitrary matrix function **a**, they wrote

$$\dot{\mathbf{a}} = \frac{2\pi i}{h} (\mathbf{Wa} - \mathbf{aW}) \, , \tag{7.81}$$

where **W** is a diagonal matrix. This was shown in Part I (see above). The elements W_n are the state energies.

BHJ then noted that (7.76–7.78), and (7.81) may be combined to give

$$\mathbf{Wq} - \mathbf{qW} = \mathbf{Hq} - \mathbf{qH} \, ,$$
$$\mathbf{Wp} - \mathbf{pW} = \mathbf{Hp} - \mathbf{pH} \, , \tag{7.82}$$

or

$$(\mathbf{W} - \mathbf{H})\,\mathbf{q} - \mathbf{q}\,(\mathbf{W} - \mathbf{H}) = \mathbf{0} \, ,$$
$$(\mathbf{W} - \mathbf{H})\,\mathbf{p} - \mathbf{p}\,(\mathbf{W} - \mathbf{H}) = \mathbf{0} \, . \tag{7.83}$$

Then $(\mathbf{W} - \mathbf{H})$ commutes with both **q** and **p**, and therefore with all functions of $(\mathbf{q}, \mathbf{p})$. Hence,

$$(\mathbf{W} - \mathbf{H})\,\mathbf{H} - \mathbf{H}\,(\mathbf{W} - \mathbf{H}) = \mathbf{0}$$

and $\dot{\mathbf{H}} = 0$.

Everything BHJ had written so far was for the coordinates and momenta $(\mathbf{q}, \mathbf{p})$. These results would also hold for another set of coordinates such as $(\mathbf{Q}, \mathbf{P})$, provided that the commutation relationship holds for both $(\mathbf{q}, \mathbf{p})$ and $(\mathbf{Q}, \mathbf{P})$. This is fully analogous with classical mechanics, where the canonical transformation between (q, p) and (Q, P) will guarantee that the canonical equations hold for each set. In quantum mechanics, this is guaranteed if, as we have seen here, the commutation relationship (7.74) holds for each set, i.e., if

$$\mathbf{pq} - \mathbf{qp} = \mathbf{PQ} - \mathbf{QP} = \frac{h}{2\pi i}\mathbf{1}\,. \tag{7.84}$$

BHJ surmised that the most general transformation that would preserve the commutation relationship would be based on a transformation matrix $\mathbf{S}$ with an inverse $\mathbf{S}^{-1}$. Indeed, if $\mathbf{P} = \mathbf{SpS}^{-1}$ and $\mathbf{Q} = \mathbf{SqS}^{-1}$, we may easily transform (7.74) to

$$\mathbf{SpS}^{-1}\mathbf{SqS}^{-1} - \mathbf{SqS}^{-1}\mathbf{SpS}^{-1} = \frac{h}{2\pi i}\mathbf{1}\,. \tag{7.85}$$

Then we have a solution for $\mathbf{P}$ and $\mathbf{Q}$ from initial values $\mathbf{p}$ and $\mathbf{q}$.

The matrix function $\mathbf{f}\,(\mathbf{pq})$ may also be transformed according to

$$\mathbf{f}\,(\mathbf{PQ}) = \mathbf{f}\left(\mathbf{SpS}^{-1}\mathbf{SqS}^{-1}\right) = \mathbf{Sf}\,(\mathbf{pq})\,\mathbf{S}^{-1}. \tag{7.86}$$

If $\mathbf{f}\,(\mathbf{pq})$ is the Hamiltonian, (7.86) implies

$$\mathbf{H}\,(\mathbf{PQ}) = \mathbf{H}\left(\mathbf{SpS}^{-1}\mathbf{SqS}^{-1}\right) = \mathbf{SH}\,(\mathbf{pq})\,\mathbf{S}^{-1}. \tag{7.87}$$

Mathematically, a Hermitian matrix can always be brought into diagonal form by means of a *unitary transformation*, that is, a transformation for which $\mathbf{S}^{-1} = \mathbf{S}^{+}$ (the adjoint, which is the conjugate of the transpose) ([128], pp. 152, 195–197). Therefore, if $\mathbf{S}$ is unitary,

$$\mathbf{SH}\,(\mathbf{pq})\,\mathbf{S}^{-1} = \mathbf{W}\,, \tag{7.88}$$

where $\mathbf{W}$ is the diagonal form of $\mathbf{H}$. The energies of the system are then known, since they are the eigenvalues of the Hamiltonian. Finding the energies essentially solves the quantum mechanical problem.

BHJ then considered a perturbation solution for more complex Hamiltonians. They chose to write the Hamiltonian as a series in λ, viz.,

$$\mathbf{H} = \mathbf{H}_0 + \lambda\mathbf{H}_1 + \lambda^2\mathbf{H}_2 + \cdots, \tag{7.89}$$

where the solution to the system with the Hamiltonian $\mathbf{H}_0$ was known. The solution matrices were $\mathbf{q}_0$ and $\mathbf{p}_0$, with

$$\mathbf{p}_0\mathbf{q}_0 - \mathbf{q}_0\mathbf{p}_0 = \frac{h}{2\pi i}\mathbf{1}\,, \tag{7.90}$$

and the Hamiltonian $\mathbf{H}_0\left(\mathbf{q}_0, \mathbf{p}_0\right)$ was diagonal. The problem was then to find a transformation $\mathbf{S}$ such that

$$\mathbf{p} = \mathbf{Sp}_0\mathbf{S}^{-1} \text{ and } \mathbf{q} = \mathbf{Sq}_0\mathbf{S}^{-1} \tag{7.91}$$

and

$$\mathbf{H}(\mathbf{q}, \mathbf{p}) = \mathbf{S}\mathbf{H}(\mathbf{q}_0, \mathbf{p}_0)\,\mathbf{S}^{-1} = \mathbf{W} \tag{7.92}$$

was diagonal. To solve for **S**, they proposed setting

$$\mathbf{S} = \mathbf{1} + \lambda\mathbf{S}_1 + \lambda^2\mathbf{S}_2 + \cdots . \tag{7.93}$$

The inverse would then be

$$\mathbf{S}^{-1} = \mathbf{1} - \lambda\mathbf{S}_1 + \lambda^2\left(\mathbf{S}_1^2 - \mathbf{S}_2\right) + \lambda^3 \cdots . \tag{7.94}$$

Collecting terms at each level of approximation, the first three diagonal matrices $\mathbf{W}_0$, $\mathbf{W}_1$, and $\mathbf{W}_2$ were therefore

$$\begin{array}{r} \mathbf{H}_0(p_0, q_0) = \mathbf{W}_0 \\ \mathbf{S}_1\mathbf{H}_0 - \mathbf{H}_0\mathbf{S}_1 + \mathbf{H}_1 = \mathbf{W}_1 \\ \mathbf{S}_2\mathbf{H}_0 - \mathbf{H}_0\mathbf{S}_2 + \mathbf{H}_0\mathbf{S}_1^2 - \mathbf{S}_1\mathbf{H}_0\mathbf{S}_1 + \mathbf{S}_1\mathbf{H}_1 - \mathbf{H}_1\mathbf{S}_1 + \mathbf{H}_2 = \mathbf{W}_2 \\ \cdots\cdots\cdots \end{array} . \tag{7.95}$$

The coordinates and momenta were also written as approximations

$$\mathbf{q} = \mathbf{q}_0 + \lambda\mathbf{q}_1 + \lambda^2\mathbf{q}_2 + \cdots , \tag{7.96}$$

$$\mathbf{p} = \mathbf{p}_0 + \lambda\mathbf{p}_1 + \lambda^2\mathbf{p}_2 + \cdots , \tag{7.97}$$

which led to

$$\begin{aligned} \mathbf{q}_1 &= \mathbf{S}_1\mathbf{q}_0 - \mathbf{q}_0\mathbf{S}_1 , \\ \mathbf{p}_1 &= \mathbf{S}_1\mathbf{p}_0 - \mathbf{p}_0\mathbf{S}_1 , \end{aligned}$$

to first order.

BHJ went on to consider time dependencies in the Hamiltonian for one dimension. In the second section they considered higher dimensional systems and in the third, eigenvalues of general Hermitian forms. Then, in the last section they considered physical applications of the theory, including the blackbody radiation problem.

7.3.5 Enter Dirac

Dirac, who was in Cambridge at the time, accepted Heisenberg's ideas and developed an entire mathematical structure to solve the problems posed.[7] ([136], p. 90). Dirac related the story of his encounter with Heisenberg's ideas in detail in the *International School of Physics "Enrico Fermi", Course LVII* at Varenna on Lake Como ([88], pp. 119–129).

At the end of August 1925, Dirac was given a copy of Heisenberg's paper by his professor Robert H. Fowler (1889–1945), who asked him what he thought of it. At first, Dirac thought it was rather complicated and of no great help, but later he read it in detail and came to realize that it was very important for what he was doing. Dirac found the difference **pq** − **qp** troubling, but he soon realized that this noncommutation was the most important part of the theory.

Dirac was very familiar with Hamiltonian mechanics, and on one of his long walks on Sunday afternoons he recalled the idea of the *Poisson bracket* (PB). On Monday morning, as soon as he could enter the library, he searched through E.T. Whittaker's treatise *Analytical Dynamics of Particles and Rigid Bodies* [288] and found that PBs were exactly what he needed to understand **pq** − **qp**.

The system trajectory in phase space (q, p) results from a canonical transformation of the coordinates and momenta, which is defined by the requirement that the system coordinates and momenta satisfy Hamilton's canonical equations at each point on the trajectory (see Sect. 7.3.4). In classical mechanics we find that for any conservative system, the PB $\{f, g\}$ of two dynamical functions f and g is a constant of the motion. The PB is defined by[8]

$$\{f, g\} \equiv \sum_{\mathrm{r}} \left(\frac{\partial f}{\partial q_{\mathrm{r}}} \frac{\partial g}{\partial p_{\mathrm{r}}} - \frac{\partial f}{\partial p_{\mathrm{r}}} \frac{\partial g}{\partial q_{\mathrm{r}}} \right) . \tag{7.98}$$

In (7.98), the subscript r identifies a coordinate or momentum of the system. Specifically, if (q, p) are the initial coordinates and momenta and (Q, P) are the final coordinates and momenta, then

$$\sum_{\mathrm{r}} \left(\frac{\partial f}{\partial q_{\mathrm{r}}} \frac{\partial g}{\partial p_{\mathrm{r}}} - \frac{\partial f}{\partial p_{\mathrm{r}}} \frac{\partial g}{\partial q_{\mathrm{r}}} \right) = \sum_{\mathrm{r}} \left(\frac{\partial f}{\partial Q_{\mathrm{r}}} \frac{\partial g}{\partial P_{\mathrm{r}}} - \frac{\partial f}{\partial P_{\mathrm{r}}} \frac{\partial g}{\partial Q_{\mathrm{r}}} \right) . \tag{7.99}$$

The solution for the dynamics of a mechanical system may then be found by a canonical transformation from the initial point to the final point. This is the basis for the simplification that Jacobi brought to Hamilton's formulation ([140], pp. 42–47 and 174–205).

[7] In the fourth edition of his *Quantum Mechanics*, Dirac writes: "Only questions about the results of experiments have a real significance and it is only such questions that theoretical physics needs to consider." ([86], p. 5).

[8] We have used curly brackets to denote the PB because square brackets are now used (universally) for the commutator (see, e.g., [189], p. 37). Dirac used $[f, g]$ for the PB of f and g.

Dirac found the connection between the commutator and the PB to be very close, indeed so close that he only needed to insert a coefficient $ih/2\pi$ in the PB to obtain the analog [83]:

$$uv - vu \text{ is the quantum analog of } \frac{ih}{2\pi}\sum_{\mathrm{r}}\left\{\frac{\partial u}{\partial q_{\mathrm{r}}}\frac{\partial v}{\partial p_{\mathrm{r}}} - \frac{\partial u}{\partial p_{\mathrm{r}}}\frac{\partial v}{\partial q_{\mathrm{r}}}\right\}. \qquad (7.100)$$

Dirac wrote a paper explaining how to obtain derivatives in the quantum theory, rather than simply presenting the idea without providing any real basis. His paper was communicated by Fowler to the Royal Society, which quickly published it. He sent a copy to Heisenberg and got a prompt and polite reply, indicating that there could be no doubt that all of Dirac's results were correct "insofar as one believes in the new theory." Heisenberg pointed out that some of Dirac's results had already been obtained in the Born and Jordan paper and the Born, Heisenberg, and Jordan paper. However, he wrote: "Your results, in particular the general definition of differentiation and the connection of the quantum conditions with the Poisson bracket, go considerably farther." Heisenberg pointed out, however, that

$$\frac{\mathrm{d}}{\mathrm{d}t}(\mathbf{pq} - \mathbf{qp}) = 0\,,$$

while Dirac had only ([88], pp. 119–125)

$$2\pi m\,(\mathbf{q\dot{q}} - \mathbf{\dot{q}q}) = 1h\,.$$

In his book *The Principles of Quantum Mechanics*, Dirac took a general approach to developing the theory. He specified and used what he called the *symbolic method*. He admitted that the preparations required to reach this method were long but, as he wrote in the introduction to the first edition of his book, it enabled one to go deeper into the nature of things. He proceeded to give a particularly elegant formulation of the analog (7.100) ([86], pp. 85–87).

Appendix E presents the exact quantum condition and the momentum operator using Dirac's symbolic method.

7.4 Summary

In this chapter we have considered a short but intense period in the history of the quantum theory. If we neglect the work on optical dispersion by Kramers and Heisenberg in Copenhagen during the last few months of 1924, all these efforts were compressed into the period between May and November of 1925. In this period an entire theoretical picture was developed primarily by three actors in Göttingen. This period includes the two weeks Heisenberg spent on Helgoland and the discussions that would have taken place when Born stopped in Tübingen and Zurich. We have told this story in some detail in Sect. 7.2, including Einstein's reaction to Heisenberg's ideas.

Heisenberg, who is the main figure in this story, was 23 years old at the time. He was already developing his relationship with Bohr's group in Copenhagen, while working on the mathematical basis of the theory in Göttingen and Munich.

We devoted much of the chapter to a detailed discussion of the actual theory that was then quantum mechanics. We began with the work carried out on optical dispersion in Copenhagen by Kramers and Heisenberg. This is important because it helped to form Heisenberg's thinking. The matrix theory was developed in the sequence of the three papers by Heisenberg, Born and Jordan, and Born, Heisenberg, and Jordan. We have kept as close as possible to the original mathematical development of the theory and we have also added a discussion of Dirac's approach. This is followed up with more details in Appendix E. Although von Neumann considered Dirac's delta function a "fiction," Dirac's formulation has become central to quantum theory ([206], pp. ix, 25, 27).

In broad terms, there are essentially two directions which can be taken in classical mechanics. One is to base the treatment on the canonical equations for the momenta and the coordinates. This was the approach taken by Born, Heisenberg, and Jordan. The second is to use the Hamilton–Jacobi equation

$$\frac{\partial S}{\partial t} + \mathcal{H}\left(\frac{\partial S}{\partial q_1}, \ldots, q_1, \ldots, t\right) = 0,$$

for the principal function S defined in (7.48). This was the approach adopted by Schrödinger.

Chapter 8
Schrödinger's Wave Theory

Gott würfelt nicht

God does not throw dice
Albert Einstein

Aber es kann doch nicht unsere Aufgabe sein, Gott vorzuschreiben, wie er die Welt regieren soll.

But it is certainly not our responsibility to tell God how to rule the universe.
Niels Bohr

8.1 Introduction

In the winter of 1925, Schrödinger derived an equation that cast the problem of determining the electron orbital energies in an atom as an eigenvalue problem. He published this in volume 79 of *Annalen der Physik* for January of 1926 as *Quantisierung als Eigenwertproblem* (Quantization as an Eigenvalue Problem). This was quickly followed by another paper in volume 79, one each in volumes 80 and 81 of the same year, and a summary in volume 28 of *Physical Review* for December of 1926 ([249]; [250]; [251]; [252]; [253]; [254]). Davisson borrowed the first of these papers from Richardson to read on his return trip to New York (see Sect. 6.5).

Schrödinger developed what we now know as the *Schrödinger equation* during a two and a half week vacation over Christmas 1925, at a villa in Arosa in the Swiss Alps. For the details surrounding this vacation we refer the reader to the book by Walter Moore (1918–2001) [200].

C. S. Helrich, *The Quantum Theory—Origins and Ideas*, History of Physics,
https://doi.org/10.1007/978-3-030-79268-8_8

For our purposes here, it is important to note that he only brought along de Broglie's thesis as his source for the physics. However, Schrödinger left no record of his thoughts while he was working on the wave theory, so we can only speculate about how de Broglie's thesis may have inspired him on the basis of what we can deduce from his first publication on the wave equation, which appeared a few weeks after his Alpine vacation [249].

In our study of Schrödinger's first paper here, the reader will note that he only considered the nonrelativistic situation. In our study of de Broglie's thesis in §6.2, however, we noted that de Broglie's ideas were based on relativistic considerations. It is logical then to ask why Schrödinger chose not to follow the relativistic path. The reason was provided by Dirac at the *International School of Physics "Enrico Fermi" Course LVII*. Dirac explained that Schrödinger was the one physicist he felt to be most similar to himself. He and Schrödinger both had a very strong appreciation for mathematical beauty. Dirac said that, after knowing him for quite a long time, Schrödinger told him that he had first been working from a relativistic point of view and obtained an equation that was a generalization of de Broglie's ideas. Unfortunately, when he first tried to apply it to the hydrogen atom, the calculation gave results that were not in agreement with experiment. Schrôdinger was very disappointed by this and put the equation aside, believing that it was no good. Later, he went back to it and noted that in the nonrelativistic case the equation gave results in agreement with experiments. This was the equation he published. In an obituary for Schrödinger, Dirac pointed out that we now know the problem was his failure to include the electron spin, which was not then known ([88], p. 136, [87], p. 356, [148], p. 258).

In this chapter we begin with the details presented by Schrödinger in his first correspondence of 1926. There he laid out the physical and mathematical basis for the approach he would undertake. We will be careful about the mathematical details, including the basis of the connection between what Schrödinger was doing and the wave descriptions of Fermat and Hamilton. We will also carry out the variational calculations.

After considering the mathematical details of the Schrödinger equation, we pick up his final publication of 1926 in the *Physical Review*. There Schrödinger presented a detailed analysis of what he understood as a new theory based on the wave description of matter. We have outlined each section of the paper in the hope that the reader will come to understand Schrödinger's thinking.

In the penultimate section of the chapter, we give voice to the group working with Born at Göttingen and we conclude with the thoughts of Bohr and Heisenberg, which form the Copenhagen interpretation.

8.2 Schrödinger's First Correspondence

In dieser Mitteilung möchte ich zunächst an dem einfachsten Fall des (nichtrelativistischen und ungestörten) Wasserstoffatoms zeigen, dass die übliche Qnantisierungsvorschrift sich durch eine andere Forderung ersetzen lässt, in der kein Wort von „ganzen Zahlen" mehr vorkommt.

In this paper I want to first show that, for the simplest case of the (nonrelativistic and undisturbed) hydrogen atom, the normal quantization rule may be replaced by another requirement, in which there is no mention of "whole numbers".

Erwin Schrödinger

Schrödinger's January 1926 paper began with the promise that the numbering system of quantum mechanics would emerge in the same natural manner in which the standing waves appear in a vibrating string. Schrödinger was promising simplicity and even the familiar. Then he began with a straightforward and beautiful deduction of the quantum wave equation from the Hamilton–Jacobi equation in the time-independent form.

For the sake of completeness, we shall begin here with the full Hamilton–Jacobi equation

$$\frac{\partial S}{\partial t} + \mathcal{H}\left(\frac{\partial S}{\partial q_1}, \ldots, q_1, \ldots, t\right) = 0, \tag{8.1}$$

which is a partial differential equation for Hamilton's principal function S in terms of the spatial coordinates $\{q_\text{i}\}$ and the time t.

The principal function S is defined by (see §7.3.3)

$$S = \int_{\text{t}=0}^{\text{t}} L\left(\{q_\text{i}\}, \{p_\text{i}\}, t\right) \text{d}t,$$

where L is the Lagrangian and $\{p_\text{i}\}$ are the momenta of the system. The Lagrangian is the kinetic energy minus the potential energy of the system. The principal function $S\left(\{q_\text{i}\}, t\right)$ is central to analytical mechanics and, although everything that can be known about the system can be obtained from $S\left(\{q_\text{i}\}, t\right)$, the principal function has no physical meaning and cannot be measured.

The basic concept Hamilton worked with had been developed by Lagrange from an idea of Maupertuis, who was cited by de Broglie in Chap. 2 of his thesis (see also §1.3.3). Hamilton's goal had been to bring to optics the same beauty, power, and harmony that Lagrange had brought to analytical mechanics. Indeed, he referred to Lagrange's work as a "kind of scientific poem" ([140], p. 38). Hamilton focused on $S\left(\{q_\text{i}\}, t\right)$, from which he could obtain the canonical momenta of the system as

$$p_\text{i} = \frac{\partial S}{\partial q_\text{i}}, \tag{8.2}$$

and the Hamiltonian as

$$\mathcal{H} = -\frac{\partial S}{\partial t}. \tag{8.3}$$

In the case of a conservative system, such as the hydrogen atom, to which Schrödinger applied his equation, the Hamiltonian is the total energy $\mathcal{E}$, which is constant. For this system, Schrödinger thus considered $\partial S/\partial t = -\mathcal{E}$ and (8.1), which becomes

$$\mathcal{H}\left(\frac{\partial S}{\partial q_1},\ldots,q_1,\ldots\right)=\mathcal{E}\,. \tag{8.4}$$

He then introduced a function ψ defined in terms of S by[1]

$$S = K \ln \psi. \tag{8.5}$$

The function ψ is then the logical equivalent of the principal function S.

Because of its definition in terms of the principal function S, ψ also has no physical meaning and cannot be measured. However, all that is knowable can be extracted from ψ. For example, the canonical momenta (8.2) are

$$p_{\mathrm{j}} = \frac{\partial S}{\partial q_{\mathrm{j}}} = \frac{\partial}{\partial q_{\mathrm{j}}} K \ln \psi = K \frac{1}{\psi}\frac{\partial \psi}{\partial q_{\mathrm{j}}}\,,$$

and (8.4) becomes

$$\mathcal{H}\left(\frac{K}{\psi}\frac{\partial \psi}{\partial q_1},\ldots,q_1,\ldots\right)=\mathcal{E}\,. \tag{8.6}$$

Writing the Hamiltonian $\mathcal{H}$ as the sum of the kinetic and potential energies, viz.,[2]

$$\mathcal{H} = \frac{1}{2m}\sum_{\mathrm{j}} p_{\mathrm{j}}^2 + V.$$

Equation (8.6) becomes

$$\left(\frac{\partial \psi}{\partial x}\right)^2 + \left(\frac{\partial \psi}{\partial y}\right)^2 + \left(\frac{\partial \psi}{\partial z}\right)^2 - \frac{2m}{K^2}\left(\mathcal{E} - V\right)\psi^2 = 0. \tag{8.7}$$

For the hydrogen atom, Schrödinger chose the Coulomb potential

$$V\left(r\right) = -e^2/r\,, \tag{8.8}$$

using Gaussian units.

Schrödinger chose not to seek a solution to an equation of the form (8.7). Rather, he presented another principle (*Forderung*) for unique, real-valued (*reelle*), twice differentiable functions ψ. He noted that, for such functions, the integral of the left-hand side (*quadratischen Form*) of (8.7) over the whole of space would be an extremum. In a footnote, he wrote rather cryptically that it had not escaped him

---

[1] This is the form of S presented in Schrödinger's *Annalen der Physik* paper of 1926 ([249], p. 361). Dirac changed this to $S = -iK \ln \psi$, without reference to Schrödinger ([86], p. 121).

[2] The Hamiltonian was originally obtained by Hamilton as a Legendre transformation of the Lagrangian. This is still the definition. For most conservative systems of interest in quantum mechanics, however, the Hamiltonian is simply the total energy ([140], p. 97).

that this formulation was not unique ([249], footnote p. 362). He, then sought an extremum of

$$J = \int \mathrm{d}\mathbf{r} \left[\left(\frac{\partial \psi}{\partial x} \right)^2 + \left(\frac{\partial \psi}{\partial y} \right)^2 + \left(\frac{\partial \psi}{\partial z} \right)^2 - \frac{2m}{K^2} \left(\mathcal{E} - V(r) \right) \psi^2 \right], \tag{8.9}$$

using the ordinary approach of the variational calculus. The designation of this integral as J, which is a functional of ψ, is Schrödinger's. Schrödinger found the condition for which the function ψ results in an extremum of (8.9) to be

$$\left(\frac{\partial^2}{\partial x^2} \psi \right) + \left(\frac{\partial^2}{\partial y^2} \psi \right) + \left(\frac{\partial^2}{\partial z^2} \psi \right) + \frac{2m}{K^2} \left(\mathcal{E} - V(r) \right) \psi = 0, \tag{8.10}$$

which is what we now know as the *Schrödinger equation*. In Appendix B, we justify the fact that the variation $\delta J = 0$ on the basis of the equivalence of Fermat's and Jacobi's principles and carry out the details of the variation of (8.9) to yield (8.10).

The rest of Schrödinger's January 1926 paper was devoted to the solution of the linear partial differential equation (8.10) for the Coulomb potential (8.8), expressed in spherical coordinates. The wave function ψ is separable into functions that each depend on just one of the spherical coordinates (r, ϑ, ϕ). The solutions for the functions of the azimuthal angle ϑ and polar angle ϕ were well known. For the solution of the radial equation, Schrödinger had the assistance of his friend, the mathematician Hermann Weyl.

When $K = h/2\pi$, the radial quantum numbers for $\mathcal{E} < 0$ provided an exact match to the Bohr energies for the hydrogen atom (see (4.15) in §4.2.2):

$$\mathcal{E}_{\mathrm{n}} = -2\pi^2 \frac{m_{\mathrm{e}} e^4}{h^2} \left(\frac{1}{n^2} \right). \tag{8.11}$$

Schrödinger's solution also provided quantum numbers for the angular momentum with dependence on the value of the principal quantum number n. This was something the Bohr model had been unable to provide.

8.3 Physical Review December 1926

In this section, we outline Schrödinger's final paper of 1926. We might suppose that he wrote this for the *Physical Review* in order to make it more accessible to a wider audience. But that would be speculation. However, the format of the paper is in numbered sections, each presenting aspects of an argument pointing to a new metaphysical position in the physical sciences, which had been emerging in his previous German language papers. The numbered sections also appear in the abstract to the original paper.

The title, *An Undulatory Theory of the Mechanics of Atoms and Molecules*, also tells us that the paper aimed to present a new theory. This was not the theory we have followed as a scientific community in our development of the present quantum theory. This was Schrödinger's wave theory. Here we shall discuss Schrödinger's paper without comment, keeping to the standard terminology of Hamilton and Jacobi, which we have clarified in footnotes.

8.3.1 An Analogy: Mechanics and Optics

Schrödinger began this section with a remark on de Broglie's idea regarding phase waves that could be associated with moving electrons or protons [74]. Here he said he would adopt the position expressed in his series of preceding papers published since the beginning of 1926 in *Annalen der Physik* ([249]; [250]; [251]; [252]; [253]), which was that material point particles consist of, or are nothing but, wave systems. This he admitted might not be correct. However, he considered that neglect of de Broglie's idea, and the treatment of only material particles, had led to serious difficulties in the mechanics of atoms that had not been resolved. So, he proposed that his point of view might not be so risky. The wave theory might also give insights into the "transitions," which remained totally mysterious.

Schrödinger then listed four advantages of his new theory:

1. The laws of motion and quantum conditions were based on the Hamiltonian principle alone.
2. The discrepancy between the frequency of motion and of emission disappeared because differences in frequency of emission coincided with differences in frequencies of motion in the wave theory. The localization of a charge in space and time was associated with a wave system, which made selection principles superfluous.
3. It seemed that the new theory would make it possible to investigate the so-called "transitions," which were still totally mysterious.
4. There were some disagreements between the old and new theories for certain energy levels and frequencies. In these cases the new theory seemed better supported by experiment.

To explain his main line of thought Schrödinger chose to consider what he called a material point of mass m moving in a potential $V(x, y, z)$. At this point he identified the principal function of Hamilton as $W(x, y, z, t)$, rather than the standard $S(x, y, z, t)$.[3] To avoid confusion, we shall use the standard

[3] We note that in Eq. (8.1) of the January 1926 *Annalen der Physik* paper, Schrödinger used the standard notation of S for the principal function (see Eq. (8.1)). He then noted that for his system the Hamiltonian was the total energy, which was constant, so that $\partial S/\partial t = -\mathcal{E}$, whence his Eq. (8.4) followed. But he did not carry out a formal separation of the principal function S.

$$S\,(x, y, z, t) = \int_{\mathrm{t_0}}^{\mathrm{t}} \mathrm{d}t\,(T - V) \tag{8.12}$$

as the principal function.[4] In (8.12), T is the kinetic energy,

$$T = \frac{1}{2}m\left(\dot{x}^2 + \dot{y}^2 + \dot{z}^2\right) = \frac{1}{2m}\left(p_x^2 + p_y^2 + p_z^2\right), \tag{8.13}$$

and V is the potential energy. In terms of $S\,(x, y, z, t)$, the momenta $\left(p_x, p_y, p_z\right)$ are

$$p_x = \frac{\partial S}{\partial x},\; p_y = \frac{\partial S}{\partial y},\; p_z = \frac{\partial S}{\partial z}. \tag{8.14}$$

Schrödinger then returned to the original form of the Hamilton–Jacobi equation, namely,

$$\frac{\partial S}{\partial t} + \frac{1}{2m}\left[\left(\frac{\partial S}{\partial x}\right)^2 + \left(\frac{\partial S}{\partial y}\right)^2 + \left(\frac{\partial S}{\partial z}\right)^2\right] + V = 0, \tag{8.15}$$

which is a partial differential equation for the principal function S. With Schrödinger's use of W for the principal function, this is Eq. (8.3) in the *Physical Review* paper. He then carried out the standard separation, which is, with our $S\,(x, y, z, t)$,

$$S\,(x, y, z, t) = S_{\mathrm{t}}(t) + W\,(x, y, z)\,. \tag{8.16}$$

Using

$$\mathcal{H} = -\frac{\partial S}{\partial t} \tag{8.17}$$

and $\mathcal{H} = \mathcal{E}$, the separation (8.16) is then

$$S\,(x, y, z, t) = -\mathcal{E}t + W\,(x, y, z)\,, \tag{8.18}$$

which, replacing S by W, is Schrödinger's (4). Using (8.18), Eq. (8.15) becomes

$$\left[\left(\frac{\partial W}{\partial x}\right)^2 + \left(\frac{\partial W}{\partial y}\right)^2 + \left(\frac{\partial W}{\partial z}\right)^2\right] = 2m\,(\mathcal{E} - V)\,. \tag{8.19}$$

The separation (8.18) results in the equality of the spatial partial derivatives of W and S, independently of which function carries the time. This is Eq. (8.4) in Schrödinger's

[4] Specifically, Schrödinger wrote

$$W = \int_{\mathrm{t_0}}^{\mathrm{t}} \mathrm{d}t\,(T - V)$$

as Eq. (8.2) in the *Physical Review* paper [254].

paper. It is also Eq. (8.1) in his first paper from January of 1926, which was for S rather than W.

Schrödinger noted here that

$$|\mathrm{grad}\ W| = [2m\,(\mathcal{E} - V)]^{1/2}\,. \tag{8.20}$$

The surfaces W are then surfaces of constant momentum.

According to Schrödinger the partial differential equation (8.19) defines a set of surfaces $W\,(x, y, z)$ in space. The numbering of one of these is completely arbitrary and may be chosen numerically by introducing an additive constant. Following Schrödinger, we may call one of these surfaces W_0. He chose to hold time constant while carrying out this labeling of the surfaces.

He then selected one side of the surface W_0 to be the positive side, by choosing the sign of grad W and removing the absolute value sign in (8.20). Then (8.20) becomes

$$\mathrm{grad}\ W = \frac{\mathrm{d}W}{\mathrm{d}n} = \sqrt{2m\,(\mathcal{E}-V)}\,, \tag{8.21}$$

where $\mathrm{d}W/\mathrm{d}n$ is the directional derivative normal to the surface W and $\mathrm{d}n$ is the differential normal to W. The totality of the points at the tips of the differentials $\mathrm{d}n$ then generate another surface $W_0 + \mathrm{d}W_0$. In this way one may construct surfaces filling the space.

Schrödinger then allowed the time to vary. He noted that, according to the separation equation (8.18), the system of surfaces $W\,(x, y, z)$ would not vary. They are functions of the spatial coordinates alone, but the presence of a non-zero contribution from $S_{\mathrm{t}}(t) = -\mathcal{E}t$ would change the values of the constants defining the surfaces W.

Since the values of the surfaces were chosen by the constants that were added, the numerical values of the surfaces then change as $S_{\mathrm{t}}\,(t)$ changes. In this picture the surfaces remain fixed with the grid lines, and their numerical values change, but we may just as well picture the grid lines as fixed and allow the surfaces of constant W to move through them, and indeed Schrödinger invites us to take the latter point of view.

As a result, the function $S_{\mathrm{t}}\,(t)$ introduces an advance in the coordinate values. We may thus transform the coordinates so that $\mathrm{d}S_{\mathrm{t}}(t)$ only represents a translation of a single coordinate q_{i}, which we choose to be perpendicular to the particular surface W of interest, during the time $\mathrm{d}t$. We also choose the orientation of the coordinates so that in the interval of time $\mathrm{d}t$ there is a positive change in the coordinate q_{i} of $\mathcal{E}\mathrm{d}t$. We then have

$$S\,(q, t + \mathrm{d}t) = W\,(q, q_{\mathrm{i}} + \mathcal{E}\mathrm{d}t)\,. \tag{8.22}$$

The W-surface then moves along the $q_{\mathrm{i}}-$axis at a velocity equal to $\mathrm{d}q_{\mathrm{i}}/\mathrm{d}t = \mathcal{E}$. That is $\mathrm{d}W/\mathrm{d}t = \mathcal{E}$, and the normal coordinate $\mathrm{d}n$ is simply $\mathrm{d}q_{\mathrm{i}}$. The chain rule applied in (8.21) then yields the phase velocity u of the wave as

$$u = \frac{\mathrm{d}n}{\mathrm{d}t} = \frac{\mathcal{E}}{\sqrt{2m\,(\mathcal{E}-V)}}. \tag{8.23}$$

This is Eq. (8.6) in Schrödinger's paper.

8.3.2 Analogy and "Undulatory" Mechanics

Schrödinger wrote that nothing he had said thus far was in any way new. However, he noted that what Hamilton did may not have been familiar to many physicists in 1926. Although the final results of Hamilton's work might have been well known, it was the pathway leading to those results that might not have been. He spoke particularly of wave surfaces, with Huygens' principle, and Fermat's principle as examples. As he pointed out, the main issue was the wavelength because one cannot use the approximation of geometrical optics unless the wavelengths concerned are much shorter than the phenomena one wishes to study.

He then noted that we had not been able to come to grips with the mechanics of the very small. This he said was related to the fact that the orbits in an atom are no longer large compared to the wavelength of the phase waves representing the "image point" of the particle. This was the case when we sought to understand orbital mechanics in an atom. The well known concepts of wave mechanics, he pointed out, were well suited to this study.

8.3.3 Significance of Wavelength

In this section, Schrödinger considered the general problem of describing the motion of an electron in an atomic orbit. He began with the wave function[5]

$$\begin{aligned} \Psi &= A\,(x, y, z)\sin\left(\frac{S}{K}\right) \\ &= A\,(x, y, z)\sin\left(-\frac{\mathcal{E}t}{K} + \frac{W\,(x, y, z)}{K}\right), \end{aligned} \tag{8.24}$$

where K has the dimensions of $[energy \times time]$. The frequency of the wave is then

$$\nu = \frac{\mathcal{E}}{2\pi K}.$$

[5] Schrödinger denoted this by ψ, which he had used in the January 1926 paper for the part of the wave function that depended only on (x, y, z). We are using $\Psi\,(x, y, z, t)$ here in order to keep ψ for the time-independent function.

Here he notes that, "one cannot resist the temptation of supposing K to be a *universal constant*, independent of the nature of the mechanical system." He chose it to be $K = h/2\pi$ so that

$$\nu = \frac{\mathcal{E}}{h}, \tag{8.25}$$

which is the Planck–Einstein photon energy. Schrödinger considered that this was obtained in an "unforced fashion."

At this point Schrödinger raised the question of the energy and the ground state relative to which it is measured. He noted that relativistic considerations eliminate this question, but preferred not to consider them in that paper. He turned rather to the wavelength, which was unambiguous and also the quantity of greatest interest. Combining the wavelength from (8.23) and (8.25), he found

$$\lambda = \frac{h}{\sqrt{2m\,(\mathcal{E}-V)}}.$$

Recalling that (8.14) provides the momenta from S by $p_{\mathrm{i}} = \partial W/\partial q_{\mathrm{i}}$, he then obtained $p = mv = \sqrt{2m\,(\mathcal{E}-V)}$. This in turn implied

$$\lambda = \frac{h}{mv}. \tag{8.26}$$

Dividing (8.26) by the size of a Keplerian atomic orbit of dimension a,

$$\frac{\lambda}{a} = \frac{h}{mva}. \tag{8.27}$$

Because the angular momentum was known to be of the order of h, he knew that λ/a was of the order of unity. He then pointed to the evident impossibility of trying to formulate a geometrical optics description if the dimensions of the problem were of the order of the wavelength of the light.

The fact that λ decreases with increasing momentum in (8.26) indicates that ordinary mechanics becomes more applicable to higher orbits. This is an example of Bohr's correspondence principle.

Schrödinger noted that, because $\mathcal{E}$ is a function of ν, (8.23) is a relation between the phase velocity u and the frequency, which is a dispersion relation. It was therefore easy to show that the velocity of the material particle was the group velocity of the waves with phase velocity u. He cited Eq. (6.11) in de Broglie's thesis, which gives the group velocity as $v = \mathrm{d}\nu/\mathrm{d}(\nu/u)$ (Schrödinger's notation).

Regarding the wave packet he was proposing to represent the material point, Schrödinger wrote: "Now it can be proved, that the motion of – let us say – the 'center of gravity' of such a parcel will, by the laws of wave propagation, follow the same orbit as the material point would by the laws of ordinary mechanics." He offered no proof. The contention, however, provides an understanding of his thinking on this issue.

8.3.4 *Application to the Hydrogen Atom*

Schrödinger began this section by announcing that he would dwell no longer on the issues of the previous section, but proceed to more interesting questions. For this he announced that he would need a wave equation.

To study the motion of a single material point in a force field, he decided to begin with the ordinary wave equation:

$$\nabla^2 \psi - \frac{\ddot{\psi}}{u^2} = 0, \tag{8.28}$$

where the double dot indicates a second time derivative. With the wave frequency given by the Planck–Einstein relation as $\nu = \mathcal{E}/h$, Eq. (8.28) restricts the time dependence of any wave disturbance to be of the form

$$\exp\left(\pm \frac{2\pi i \mathcal{E} t}{h}\right). \tag{8.29}$$

Introducing his expression (8.23) for the phase velocity into (8.28), Schrödinger obtained

$$\nabla^2 \psi - 2m\,(\mathcal{E} - V)\,\frac{\ddot{\psi}}{\mathcal{E}^2} = 0. \tag{8.30}$$

Using $\ddot{\psi} = -\left(4\pi^2 \mathcal{E}^2/h^2\right)\psi$, the Eq. (8.30) becomes

$$\nabla^2 \psi + \frac{8\pi^2 m}{\hbar^2}\,(\mathcal{E} - V)\,\psi = 0. \tag{8.31}$$

For the potential energy, he used the Coulomb energy $V = -e^2/r$ for a single electron moving in the region around a single proton. This gave him the equation

$$\nabla^2 \psi + \frac{8\pi^2 m}{\hbar^2}\left(\mathcal{E} + \frac{e^2}{r}\right)\psi = 0, \tag{8.32}$$

which is Eq. (8.16) in Schrödinger's paper.

Schrödinger notes that the only solutions that are continuous, finite, and single-valued throughout the whole of space are those for which

$$\mathcal{E} > 0, \tag{8.33}$$

or those for which

$$\mathcal{E} = -\frac{2\pi^2 m e^4}{h^2 n^2} \text{ for } n = 1, 2, 3, \ldots . \tag{8.34}$$

The energies $\mathcal{E} > 0$ correspond to the standard hyperbolic orbits, and the negative energies were those of the Bohr orbits, which he referred to as elliptic. He noted that the hyperbolic orbits were not subject to quantization.

According to Schrödinger, the solution for negative energies involves the standard spherical harmonics and a radial function that decreases rapidly with distance from the origin, so that the size of the hydrogen atom becomes discernible. He then described the statistical weights of the levels with $n > 1$, indicating that these were degenerate, consisting of a number of states of the same energy. In these cases, the theory gave the correct number states.

8.3.5 Discrete Characteristic Frequencies

In this section, Schrödinger went to some lengths to explain why (8.32) only yields solutions for the restricted energies (8.34). The bound solutions with $\mathcal{E} < 0$ were of primary interest because they were the values Bohr obtained for the energies of the bound states through completely different reasoning. Providing a sufficient basis for these values was a critical issue in this paper. He had no intention of providing the complete mathematical proof, which in the previous section he called "tedious." But he knew his readers would require some justification.

Schrödinger separated (8.32) in the standard fashion for the spherical case as a product of functions of (ϑ, ϕ) and (r). The angular functions were the spherical harmonics. The radial equation was something new to him. In his notation, with $n\,(n+1)$ resulting from the spherical harmonics and $K = h/2\pi$, this radial equation was ([249], p. 363)

$$\frac{\mathrm{d}^2\chi}{\mathrm{d}r^2} + \frac{2}{r}\frac{\mathrm{d}\chi}{\mathrm{d}r} + \left[\frac{2m\mathcal{E}}{K^2} + \frac{2me^2}{K^2 r} - \frac{n\,(n+1)}{r^2}\right]\chi = 0\,, \tag{8.35}$$

with the requirement

$$n = 0, 1, 2, 3, \ldots\,.$$

In a footnote, he extended his warmest thanks to his friend the mathematician Hermann Weyl for introducing him to a method for dealing with this equation. This solution is now the one that appears in almost every text on quantum mechanics (e.g. [189], pp. 264–267; [82], pp. 158–162]).

Schrödinger accepted that the mathematics he had used might not be familiar to his readers. If this had been a boundary value problem, he suggested that all physicists would have been familiar with what he called "characteristic values" and "characteristic functions."[6] But there were no boundaries here. Equation (8.35) had singular points at both the origin and infinity, resulting in two different solutions.

[6] These terms are the English translation of the now familiar German terms "eigenvalue" and "eigenfunction".

These became a single solution for the values of $\mathcal{E}$ in (8.34). Schrödinger was quite certain that this would not be a familiar approach to his readers.

8.3.6 *Intensity of Emitted Light*

Schrödinger now turned to some results that he had obtained from this new mechanics.

He first noted the states of the harmonic oscillator as

$$\left(n+\frac{1}{2}\right)h\nu_0 \text{ with } n=0,1,2,3,\ldots,$$

rather than the $nh\nu_0$ of the ordinary quantum theory. The states of the moment of inertia I of a rigid rotator were

$$n\,(n+1)\,\frac{h^2}{8\pi^2 I}$$

in the new theory, replacing what he referred to as the well known $n^2h^2/8\pi^2I$ based on the old theory. He pointed out that these corrections were not small, but due to the complexity of the general solutions to (8.31), it would hardly have been possible to determine them by direct methods. A perturbation method that he and Mr. Fues of the Rockefeller Institution developed, Schrödinger said, had overcome many of these difficulties.

He also analyzed the Stark effect in which an electric field was applied, resulting in the splitting each of the Balmer (Bohr) energies (8.34) into n^2 lines. This he pointed out was because the electric field broke the symmetry of the Coulomb field. These results coincided with Epstein's.

Schrödinger admitted that he had not attached any definite physical meaning to the wave function ψ. He promised, however, that in §8 he would discuss a certain electrodynamical meaning for ψ. This would convert the atom

> [...] into a system of fluctuating charges spread out in space, and generating a specific electrical moment that changes with time with a superposition of frequencies, which exactly coincide with the differences in the vibration frequencies $\mathcal{E}/h$, i.e., coincide with the frequencies of the emitted light.

He ended this section with a positive comment on the agreement between the low-field results he had obtained for the Stark effect, for which the experimental field was 100,000 V/cm. He emphasized that he had been

> [...] led to the foregoing nearly classical calculation of the intensities by noticing *a posteriori*, i.e., after the main features of the undulatory theory had been developed, its complete mathematical agreement with the theory of matrices put forward by Heisenberg, Born, and Jordan.

([135], [23], [24], [25], [134], [220], [83], [84], [85]) The connection, he indicated,

> [...] is a rather intricate one and is by no means to be observed at first sight.

8.3.7 Wave Equation from a Variational Principle

At the beginning of this paper, Schrödinger noted that he had required Hamilton's principle to be the basis of the wave theory. This included both the laws of motion and the quantum conditions. To prove this, he had to show that the fundamental equation (8.31) with which he had been working could be obtained from an integral variational principle. He used a form of the Hamilton–Jacobi equation on which he had based the first presentation of his theory in January of 1926. He began the discussion with the problem of finding the extremum of the integral

$$I_1 = \int \mathrm{d}x\mathrm{d}y\mathrm{d}z \left\{ \frac{\hbar^2}{2m} \left[\left(\frac{\partial\psi}{\partial x}\right)^2 + \left(\frac{\partial\psi}{\partial y}\right)^2 + \left(\frac{\partial\psi}{\partial z}\right)^2 \right] + V\psi^2 \right\}, \tag{8.36}$$

subject to the normalization condition

$$I_2 = \int \mathrm{d}x\mathrm{d}y\mathrm{d}z\psi^2 = 1 \tag{8.37}$$

on the function ψ.

He then introduced (8.37) into the variational problem and set out to apply Lagrange's method of undetermined multipliers. For this problem, he indicated that the multiplier would be the energy $-\mathcal{E}$ and the resulting problem would be to find the variation of

$$I = \int \mathrm{d}x\mathrm{d}y\mathrm{d}z \left\{ \frac{\hbar^2}{2m} \left[\left(\frac{\partial\psi}{\partial x}\right)^2 + \left(\frac{\partial\psi}{\partial y}\right)^2 + \left(\frac{\partial\psi}{\partial z}\right)^2 \right] + (V - \mathcal{E})\,\psi^2 \right\}. \tag{8.38}$$

This was the same mathematical problem that appeared in the January 1926 paper (see (8.9)). The integrand in (8.36) would simply be the Hamiltonian of the system if one identified $p_x = (h/2\pi)\,\partial\psi/\partial x$, $p_y = (h/2\pi)\,\partial\psi/\partial y$, and $p_z = (h/2\pi)\,\partial\psi/\partial z$. Before performing the variation, he reformulated it as a general mechanical problem for which he chose the Hamiltonian to be

$$\mathcal{H}_{\mathrm{gen}} = \frac{1}{2m} \sum_{\mathrm{j,k}}^{\mathrm{N}} a_{\mathrm{jk}} p_{\mathrm{j}} p_{\mathrm{k}} + V, \tag{8.39}$$

where $p_1, \ldots p_{\mathrm{N}}$ would be replaced by $p_1 = (h/2\pi)\,\partial\psi/\partial q_1, \ldots p_{\mathrm{N}} = (h/2\pi)\,\partial\psi/\partial q_{\mathrm{N}}$. The general quadratic form for the integrand in (8.38) was thus

$$\frac{(h/2\pi)^2}{2m}\sum_{\mathrm{j,k}}^{\mathrm{N}} a_{\mathrm{jk}}\frac{\partial\psi}{\partial q_{\mathrm{j}}}\frac{\partial\psi}{\partial q_{\mathrm{k}}} - (\mathcal{E} - V(\mathbf{q}))\,\psi^2, \tag{8.40}$$

where we have used the notation $q = q_1, \cdots q_{\mathrm{N}}$. The integral (8.38) then becomes

$$I_{\mathrm{gen}} = \int \mathbf{dq}\left[\frac{(h/2\pi)^2}{2m}\sum_{\mathrm{j,k}}^{\mathrm{N}} a_{\mathrm{jk}}\frac{\partial\psi}{\partial q_{\mathrm{j}}}\frac{\partial\psi}{\partial q_{\mathrm{k}}} - (\mathcal{E} - V(r))\,\psi^2\right], \tag{8.41}$$

where $\mathbf{dq} = \mathrm{d}q_1 \cdots \mathrm{d}q_{\mathrm{N}}$.[7]

The details of the variation $\delta I_{\mathrm{gen}} = 0$ have been carried out in Appendix B. The result is

$$\frac{\hbar^2}{2m}\sum_{\mathrm{j,k}}^{\mathrm{N}} a_{\lambda\mu}\frac{\partial^2\psi}{\partial q_{\mathrm{j}}\partial q_{\mathrm{k}}} + (\mathcal{E} - V(q))\,\psi = 0\,. \tag{8.42}$$

This is Eq. (8.26) at the end of §7.

8.3.8 *Physical Meaning of the Wave Equation*

Schrödinger began this section by referring to the remark he had made in §6 regarding the real physical meaning of the wave function ψ. He had put this off until he was prepared to treat a general problem in a wholly arbitrary system. Equation (8.42) was precisely (the generalization of) the time-independent form of the Schrödinger equation that he needed.

He now introduced the time into Eq. (8.42). Up to this point all his results had been based on the assumption that the time dependence of ψ was only through the factor $\exp(\pm 2\pi i\mathcal{E}t/\hbar)$ (see (8.29)). He included the $\pm$ at this point, noting that the sign on the argument of the exponential makes no difference because the meaning of ψ is only in the product of ψ and its complex conjugate. Then,

$$\frac{\partial\psi}{\partial t} = \pm\frac{2\pi i\mathcal{E}}{h}\psi. \tag{8.43}$$

For a particular characteristic value (eigenvalue) of the energy $\mathcal{E}_{\mathrm{j}}$, this yields

$$\exp\left[\left(\frac{2\pi\mathcal{E}_{\mathrm{j}}}{h}t + \theta_{\mathrm{j}}\right)i\right],$$

[7] If we consider a transformation of coordinates, there must be a Jacobian for the transformation ([138], pp. 378–79). However, we are beginning our development from the generalized coordinates **q**, rather than transforming coordinates.

where θ_j is a phase constant. He also included here the definition $i = \sqrt{-1}$.[8]

If the characteristic functions (eigenfunctions) corresponding to $\mathcal{E}_j$ were u_j, the general solution was[9]

$$\psi = \sum_{j=1}^{\infty} c_j u_j \exp\left[\left(\frac{2\pi i \mathcal{E}_j}{h} t + i\theta_j\right)\right].$$

Parenthetically, Schrödinger also assumed at this point that all the characteristic values (eigenvalues) had to be discrete. In addition, he had chosen the coefficients c_j to be real. Denoting the complex conjugate by an overbar, the square of the absolute value of the complex function ψ is

$$\psi\bar{\psi} = 2\sum_{j,j'=1}^{\infty} c_j c_{j'} u_j u_{j'} \cos\left(\frac{2\pi\left(\mathcal{E}_j - E_{j'}\right)}{h} t + \theta_j - \theta_{j'}\right).$$

He pointed out that a difficulty might result in attaching physical meaning to the wave function, since the coordinates involved were the generalized coordinates rather than ordinary spatial coordinates. The difficulty, however, disappears in the case of the hydrogen atom. This allowed a fairly accurate calculation of the results of the Stark effect experiment, as he had noted earlier. To accomplish this he had made the hypothesis

> [...] that the charge of the electron is not concentrated in a point, but is spread out through the whole space proportional to the quantity $\psi\bar{\psi}$.

He cautioned his reader that we must still be aware that the electron is restricted to a region with linear dimension a few angstroms, since the wave function ψ practically vanishes outside of a small region near the nucleus (see §4). To calculate the radiation from the atom, he then needed only to calculate the rectangular components of the total electric moment. For example, the z component would be

$$c_j c_{j'} \int\int\int \mathrm{d}x\mathrm{d}y\mathrm{d}z\,(z)\,\psi\bar{\psi}. \tag{8.44}$$

The frequency of the emitted radiation would be $\left(\mathcal{E}_j - E_{j'}\right)/h$ and the intensity would be proportional to the square of (8.44). The results of these calculations were in fair agreement with experiment.

The triple integral (8.44), Schrödinger noted, was what the Heisenberg theory would call an element of the matrix $z\left(j, j'\right)$. This, he said, was the intimate connection between the undulatory and matrix theories.

---

[8] In 1926 Schrödinger was very careful about possible differences in the mathematical backgrounds of his readers.

[9] At the time Schrödinger thought the characteristic (eigen) functions had to be real functions. He made that point in the January 1926 paper.

Schrödinger considered that the greatest achievement of the new theory was that it could, by the localization of the atomic charge in space and time, compute both the frequencies and intensities of the emitted light using ordinary electrodynamics. The selection rules then automatically resulted from the vanishing of the triple integral in (8.44), although he admitted that Heisenberg's formal theory would be more helpful in treating N-electron atoms. Then the single electron charge would have to be replaced by a sum $\sum e_i z_i$ and the integral would be over coordinates $(x_1, y_1, z_1, \ \ldots \ x_N, y_N, z_N)$.

8.3.9 Non-conservative Systems and Dispersion

Using (8.43), Eq. (8.42) becomes

$$\frac{\hbar^2}{2m}\sum_{j,k}^{N} a_{\lambda\mu}\frac{\partial^2\psi}{\partial q_j \partial q_k} - V\psi \mp i\hbar\frac{\partial\psi}{\partial t} = 0. \tag{8.45}$$

This is the full time- and space-dependent Schrödinger equation. According to Schrödinger, one can easily make the potential non-conservative by adding in a time-dependent term, such as the one resulting from an incident light wave. He chose not to go into detail, but noted:

> The effect of the incident light wave is that two forced vibrations with in general very small amplitudes, and with the two frequencies $E_j/h \pm \nu$, are associated with each free vibration of the undisturbed system with frequency E_j/h, where ν is the frequency of the incident light.

He identified these forced frequencies with the

> [...] secondary wavelets that are necessary to account for the absorption, dispersion, and scattering.

He also found that the amplitude of these secondary peaks increased significantly as the frequency ν approached a natural emission frequency $\left(E_j - E_k\right)/h$ of the atom. The final formula is almost identical to the Helmholtz dispersion formula developed by Kramers ([162], [163]).

The present theory, he admitted, had no damping term to prevent the classical infinite amplitude at resonance. Nevertheless, a resonance with infinite magnitude was not encountered. The amplitude of the forced vibration simply attained a maximum. The amplitude of the original free vibration was also decreased. He wrote:

> This behavior seems to afford an insight (though incomplete) into the so-called transition from one stationary state to another which hitherto has been totally inaccessible to computation.

8.3.10 Relativity and the Magnetic Field

In this last section of the paper, Schrödinger noted that he had dealt neither with relativity nor with magnetic field effects, admitting that this might seem strange since these were central to de Broglie's thesis, which was the original source of the ideas leading to the theory. He also pointed out that his ideas could be taken some way toward including relativity and magnetic fields, such as would be required to study the Zeeman effect. However, Schrödinger gave two reasons that made him limit his choice.

The first of these was the fact that relativistic theory had not, at that time, been able to treat more than the single-electron hydrogen atom. However, there were still many interesting problems that the new theory could consider beyond those involving a single electron.

The second was the fact that the relativistic theory of the hydrogen atom was apparently incomplete. Schrödinger was concerned about the appearance of half integer quantum numbers in Sommerfeld's theory of the fine structure components in hydrogen. He indicated that this might be related in some way to the George Uhlenbeck (1900–1988) and Samuel Goudsmit (1902–1978) theory of the spinning electron, but he had no more specific ideas about this ([274], [275]).

8.4 Summary

In this chapter we have chosen to follow Schrödinger to the letter. We have presented the development of the Schrödinger equation including some of the mathematical details not present in the published papers. In §A.2, we presented the basis for Schrödinger's variation of the spatial integral which legitimized Schrödinger's linear equation. From that starting point, his solution of the hydrogen atom follows from the standard separation of variables, common to our present-day courses on quantum mechanics. Here Schrödinger gave specific thanks to Weyl for the solution for the radial component, which yielded the Bohr energies. Also noteworthy is the question of the complex variable in the equation defining the wave function ψ in terms of the principal function S.

We devoted considerable space to Schrödinger's last paper of 1926, in which he presented some rather detailed discussions of his interpretation of the physics and his understanding of the meaning of $|\psi|^2$. These are interesting because they provide details on Schrödinger's conception of his theory which are not normally discussed in detail in a course on quantum mechanics. In modern teaching of the subject, however, Schrödinger's thoughts are often referred to as different from our present understanding, but students are not encouraged to ask for an explanation.

If we ignore Schrödinger's belief in the wave basis of matter, we may progress comfortably with our teaching of quantum mechanics by following Dirac's formulation, without concerning ourselves or our students with Schrödinger's world view.

Chapter 9
Spin and Its Interpretation

[…] even today the physicist more often has a kind of faith in the correctness of the new principles, rather than a clear understanding of them.

Werner Heisenberg

9.1 Introduction

The proposal for the spin of the electron preceded Schrödinger's development chronologically. He referred to this in the last section of his *Physical Review* paper of 1926. Because of the significance of this proposal, as well as the unusual nature of the story, it is important to spend a little time on the issue of spin.

The last two sections of the present chapter both deal with responses to Schrödinger's equation. The physics community did understand the importance of the work Schrödinger had done, although it did not follow him on the path he was proposing. There was serious disagreement from both the group in Göttingen and the one in Copenhagen. An interesting accident of history, we may say, resulted from the fact that Heisenberg was a member of both of those groups and had worked on the foundations of matrix mechanics.

The main statement from Göttingen that dealt directly with Schrödinger's equation was Born's interpretation of $|\psi|^2$, which is now widely accepted. The interpretation from Copenhagen is important from a historical point of view, whether or not the reader chooses to accept it. There can be no real objection to Heisenberg's indeterminacy principle on a mathematical basis. We may wonder at some of the intricacies inherent in Bohr's thinking, but the basic importance of experiment is clear. Note, however, that the key experiments are no longer normally desktop experiments.

We shall end our study before we actually need to pick up the Bohr–Einstein debates in any detail.

C. S. Helrich, *The Quantum Theory—Origins and Ideas*, History of Physics,
https://doi.org/10.1007/978-3-030-79268-8_9

9.2 Spin of the Electron

Sommerfeld's former student, Pauli, after spending time at Göttingen and Copenhagen, accepted a position at the University of Hamburg in 1923. Possibly as a result of a teaching assignment, Pauli became concerned with the lack of understanding of the basis of the periodic table, which had to result from the arrangements of the electrons in atomic orbits. These arrangements were determined by assigning a set of three classically-based quantum numbers to each electron. Toward the end of 1924, it occurred to Pauli that the task of finding an appropriate arrangement of the electrons might be improved by adding a fourth, non-classical quantum number to each electron. This fourth quantum number would have two values. He soon realized that his proposal could lead to a resolution of the problem of closed orbits. In January of 1925, Pauli announced his *exclusion principle*, which was that no two electrons in a single atom could have the same four quantum numbers [3].

In 1925, inspired by Pauli's proposal, Goudsmit and Uhlenbeck, two graduate students of Ehrenfest's at Leiden University in the Netherlands, suggested that the fourth quantum number might be the electron spin. When Goudsmit and Uhlenbeck explained this idea to Ehrenfest, he was interested. Uhlenbeck wrote that Ehrenfest thought the idea was either very important or nonsense, but that it should at any rate be published, and he asked Goudsmit and Uhlenbeck to write a short, modest letter to *die Naturwissenschaften* and give it to him. Then he suggested that they could discuss the idea with Lorentz.

Lorentz had retired, but gave a lecture every Monday morning at 11:00. Uhlenbeck said that everyone who could possibly be there was present. Lorentz was kind and interested, wrote Uhlenbeck, and returned the following Monday with a stack of papers filled with careful calculations showing that the proposal would require the velocity of the surface of the electron to be ten times the speed of light. When Uhlenbeck brought this information to Ehrenfest, suggesting that their letter should not be published, Ehrenfest told him that the letter had been sent quite a while ago and would soon be published. And then he added:

> Sie sind beide jung genug um sich eine Dummheit leisten zu können!

(You are both young enough to be able to afford a little stupidity!) ([273], [279]). The basis of the structure of atoms was beginning to fall into place with the help of the quantum theory, even though mathematical details could not be obtained.

Figure 9.1 is a photograph of the sixteenth century Academy Building of Leiden University on the Rapenburg canal, Netherlands.

9.3 Interpretation of the Wave Function

Toward the end of the summer semester of 1926, Arnold Sommerfeld invited Schrödinger to present his ideas at a seminar in Munich. Heisenberg was visiting his

Fig. 9.1 The sixteenth century Academy Building of Leiden University on the Rapenburg canal, Netherlands (*Author* Rudolphous)

parents in Munich at that time and was able to attend Schrödinger's presentation. He was enthralled by the mathematical presentation.

However, as Schrödinger began to describe the details of what seemed to be going on in the wave picture as the electron transitioned from one stationary state to the next, Heisenberg thought he had gone too far and he raised an objection. Wien responded rather sharply to Heisenberg, saying that he understood Heisenberg's difficulty, but assured him that the quantum mechanics of Göttingen was now at an end and that Schrödinger would soon put aside all the difficulties and the nonsensical quantum jumps. In his own response, Schrödinger was not so confident that all would be accomplished in such a short time, although he remained convinced of the power of his wave theory.

That evening Heisenberg wrote to Bohr, describing the lecture and the exchange that had occurred. Bohr then invited Schrödinger to come to Copenhagen for two weeks in September to discuss his theory. Those two weeks have been treated in some detail in many places (cf. [238], pp. 129–131), although the source is always Heisenberg's description ([136], pp. 105–109). The discussions did not go well. Heisenberg pointed out that, rather than his usual courteous and accommodating self, Bohr was like a fanatic who would give no ground. Heisenberg devoted three pages of his book *Der Teil und das Ganze* to examples of the discussions.

After Schrödinger had left, Bohr often came late at night to Heisenberg's room under the roof of the institute where they discussed possible thought experiments, often until long past midnight. They wanted to make sure that they completely understood the theory. It was soon clear that Bohr and Heisenberg were seeking solutions to the difficulties in different directions.

Bohr wanted to have the concepts of particle and wave stand together in a complete picture of reality. Heisenberg's position was that the theoretical structure of the quantum theory provided a definite physical interpretation for each of the calculable quantities, such as energy, momentum, and frequencies. Bohr and Heisenberg normally came to the same understanding late in the night. But the discussions went on without complete agreement until Bohr went to Norway for a skiing vacation in February of 1927 ([136], pp. 109–110).

After Bohr's departure, Heisenberg found himself alone to pursue the questions that interested him. These included the issue of the trajectory of an electron as recorded in a cloud chamber. In a cloud chamber, we do not actually observe the trajectory of an electron. What we observe is the collection of droplets resulting from the electron's collisions, and these are much larger than the electron. Therefore, the position of the electron along the trajectory was not actually known, even as a result of measurements.

Mathematically, the trajectory of the electron is a combination of its position and momentum. How well the trajectory is known is then a question of how quantum mechanics deals with the momentum of the electron if the position is known to within a certain undetermined limit.[1] Heisenberg found that a brief calculation gave him an answer to this question: the product of the indeterminacies of the position and the momentum of a particle cannot be smaller than Planck's constant. That is $\Delta p \Delta q > h^2$ ([136], p. 112).

Bohr returned from his skiing vacation with a solution to the problem of the dual nature of the electron. He proposed that, although the electron possessed both wave and particle natures, these were *complementary* and could not be observed together at the same instant.[3] At first Bohr had reservations about Heisenberg's *indeterminacy principle*. But the Swedish physicist Oskar Klein (1894–1957), who was then working at Copenhagen, was helpful in pointing out that there was, in fact, no real difference between Bohr's ideas and Heisenberg's indeterminacy principle. Indeed, Klein suggested that Bohr's ideas might become more understandable to the physics community through Heisenberg's mathematics[4] ([136], p. 113).

The discussions between Bohr and Heisenberg were nevertheless difficult. Heisenberg admitted this, although he devoted very little space to the discussions with Bohr before acknowledging, with gratitude, the efforts of Klein ([136], p. 113). It is also clear from the historical record that Heisenberg became a supporter of Bohr's position

---

[1] Heisenberg writes that this approach came to him as he remembered what Einstein had told him about measurement during their discussion in Einstein's apartment in Berlin: "*Erst die Theorie entscheidet darüber, was man beobachten kann.*"

First the theory determines what one can observe ([136], p. 111).

[2] The result is actually $\Delta q \Delta p \geq \hbar/2$ where $\hbar = h/(2\pi)$.

[3] The word "complementary" is derived from the Latin *complementum*, which refers to "that which completes" ([238], p. 131). The quote from Schiller

Nur die Fülle führt zur Klarheit
Und im Abgrund wohnt die Wahrheit.
Only wholeness leads to clarity,
And truth lives in the background

was often used by Bohr ([238], p. 60).

[4] The agreement between Bohr and Heisenberg was not as easily reached as Heisenberg indicates. There was, however, complete agreement at the time of the Solvay Conference in Brussels in the Autumn of 1927 [1].

([290], p. 164). We see this in the preface to Heisenberg's book *The Physical Principles of the Quantum Theory*, based on the lectures he delivered at the University of Chicago in 1929 [132]. There he wrote:

> The purpose of this book seems to me to be fulfilled if it contributes somewhat to the diffusion of that "*Kopenhagener Geist der Quantentheorie* (Copenhagen spirit of the quantum theory)," if I may so express myself, which has directed the entire development of modern atomic physics.

Schrödinger's interpretation of the wave function was unacceptable to the group at Göttingen. There seemed to be considerable evidence of the particle nature of matter, such as clicks on the Geiger counter and tracks in a cloud chamber, which indicated a locality that was not the property of a wave. Born, however, acknowledged that he readily took up Schrödinger's approach because the ψ-function held the promise of interpretation. He was thinking of what Einstein had done to make the idea of the photon more comprehensible. In his paper *On a heuristic point of view concerning the production and transformation of light*, Einstein had interpreted the wave intensity (the square of the wave amplitude) in terms of the probability for the occurrence of photons. Born reasoned that this concept could be carried over directly to the ψ-function if $|\psi|^2$ could be shown to represent the probability density of electrons (or other particles).

To establish this, Born considered the collision of an electron with a heavy atom, which, because of its relatively high mass, could be considered to remain stationary. The atom, and the probabilities of possible transitions among atomic levels, were described by the Schrödinger equation. The electron colliding with the atom was considered to be part of a beam of electrons described by a Schrödinger plane wave with intensity $|\psi_0|^2$. The spatial region in which the electron was considered to interact with the atom was defined by a spatially dependent potential energy. A plane wave of diminished intensity, representing the electrons that did not interact with the target atom, passed through this region of interaction and on to infinity.

Part of the incoming electron wave, representing the electrons that interacted with the atom, was transformed into a secondary spherical wave, ψ_1, which emerged from the stationary site of the atom. At great distances from the atom, the square magnitude of this receding spherical wave $|\psi_1|^2$ depended on the direction from the atom. This $|\psi_1|^2$ was a measure of the probability that a colliding electron left the site of the collision in that direction.

Born pictured this as the waves of a passing ship, the electron beam, colliding with a pile supporting a wharf, the atom, which produced secondary waves emerging from the pile, the waves of ψ_1. There was a fundamental disagreement between this understanding of the wave function and the one Schrödinger had proposed in §8.3.8 ([28]; [190], Chap. XIX). However, as an important check on Born's idea, Gregor Wentzel (1898–1978) was able to derive Rutherford's results for the scattering of α-particles from Born's result [280].

Born noted, however, that Heisenberg's publication containing the indeterminacy principle[5] did far more than anything else to bring about acceptance of the statistical

[5] The term Heisenberg used for

interpretation of the wave function. The issue was that the electron did not behave like a classical point particle which would possess a position and a momentum. As the accuracy with which we know the position of the electron increases, the accuracy with which we know the momentum decreases proportionately, and vice versa [28], [133].

Born received the 1954 Nobel Prize in Physics "for his fundamental research in quantum mechanics, especially for his statistical interpretation of the wavefunction."

9.4 Copenhagen Interpretation

The basic understanding reached by Bohr, Heisenberg, and Klein, including complementarity and the role of indeterminism, forms the basis of what is known as the *Copenhagen interpretation* of the quantum theory. The presentation of the Copenhagen interpretation by some authors is rather short, as in [189] and [126], or missing entirely, as in [190]. Andrew Whitaker (b. 1946) devotes an entire chapter to the ideas from Copenhagen [288]. The choice depends on the objectives of the author and Whitaker's topic is the Bohr–Einstein debates. Here we will consider the development of the Copenhagen interpretation because it is an integral part of the history of quantum mechanics.

In 1957 Léon Rosenfeld (1904–1974), a long-term collaborator of Bohr's, pointed out that interpreting a formalism is a false problem. In a good theory, ordinary language, with some added technical terms, is inseparably united with any necessary mathematical apparatus ([290], p. 158). Rudolf Peierls (1907–1995) also pointed out that if we use the term "Copenhagen interpretation" we are implying that there is more than one interpretation. In 1986 he wrote ([290], p. 159):

> [...] when you refer to the Copenhagen interpretation of the mechanics what you really mean is quantum mechanics.

But these definitive statements were written long after quantum mechanics had become the standard language in physics.

We have already seen that there was already a problem for anyone who understood what Planck was saying when he spoke to the December 1900 meeting of the DPG in Berlin. From our consideration of Planck's thoughts in the years following that presentation, we realize that he was deeply troubled by the quantum. It did not fit into the electromagnetic wave theory of Maxwell. And Einstein's heuristic ideas about the light particle did nothing to rescue Maxwell's theory.

The emerging quantum ideas had now matured into a theory with what seemed to be a solid mathematical basis. In the spirit of Einstein, Heisenberg had asked the

---

$$\Delta p \Delta q \geq \frac{1}{2}\left(\frac{h}{2\pi}\right)$$

was *Unbestimmtheitsrelation*. Unbestimmtheit can be translated as either indeterminism or uncertainty. I have used indeterminism because this seems closest to Heisenberg's mathematical approach. Both translations are acceptable.

theory what could be observed and the theory had responded. The response had not been quite what anyone had anticipated, but it was elegant.

In our treatment of Heisenberg's principle in Appendix B, we have tried to make the argument as objective as possible. Our discussion only requires general mathematical operators, which satisfy a particular commutation relation that is known to hold for the position and momentum. We also provide a separate derivation of the Schwarz inequality to remove any difficulties. In this form, Heisenberg's indeterminacy has mathematical elegance. Placed in the context of Bohr's complementarity, it takes on the role of a principle of physics.

It is very difficult to insist that physicists, whether experimentalists or theoreticians, have no mental picture of the quantum particle or energy parcel. Bohr was aware of this as well as the limitations of human language to describe the energy parcel. Heisenberg told an anecdote that captures part of Bohr's thinking on this issue. While washing dishes on a skiing trip, Bohr noticed how dirty the water and wash towels appeared to be, while the dishes, nevertheless, became clean. He remarked that this was like our attempts to clarify quantum phenomena in terms of human language. In spite of the limitations of language, we were still able to bring clarity and understanding ([136], p. 190).

On the surface, Bohr's approach to the dilemma presented by the particle or wave concept of a quantum particle (the energy parcel) was of a different nature than Heisenberg's approach to the measurement of a trajectory. Beyond the mathematically equivalent wave mechanics of Schrödinger and matrix mechanics of Born, Heisenberg, and Jordan, there was a difference in the behavior of the electron depending on the experiment performed. Davisson and Germer had used the crystal lattice of nickel to observe the electron and found it to be a wave, while the tracks in a cloud chamber showed the electron to be a particle.

The validity of neither experiment could be questioned. Bohr's idea was that the measurements in these two experiments were complementary. The actual experiment carried out was crucial to the behavior of the electron. Indeed, it was a wave when it reflected from the closely spaced planes of the nickel lattice, whereafter the diffracted electrons were finally collected in a Faraday box and recorded with a galvanometer; but it was a particle when it left the site of a Compton collision and collided with the alcohol or water molecules in the supersaturated solution contained in the cloud chamber, where the resulting condensed droplets could be photographed. These two experiments were completely different and the corresponding behavior was, too.

In each case the results were easily understood in terms of classical physics. Bragg's law for wave (X-ray) diffraction is a law of classical physics. And the motion of the electron between collisions in the cloud chamber is understood in terms of Hamilton's canonical equations, which are an expression of Newtonian mechanics. Finally, we can describe each experiment very well in terms of ordinary language.

These were the points that Bohr was making regarding quantum measurements:

1. The quantum property was always measured in an experiment. The experiment must be described in detail, including the location and function of the instruments.
2. All instruments must be classical, providing data in classically understood terms.

3. The language we use to describe the experiments is ordinary human language, which was not developed to handle quantum phenomena.
4. There will be a separation point between the quantum particle being measured and the instrument performing the measurement. We cannot identify this point precisely, because if we could, we would be able to identify the quantum particle and its properties. This separation point is called the *Heisenberg cut* ([290], p. 172). The location of the Heisenberg cut is arbitrary, but it must be at a point where classical physics assuredly holds.

Henrik Zinkernagel devotes considerable space to a detailed description of Bohr's insistence that all experimental measurements must be carried out with classical instruments [295]. Whitaker notes the presence of this idea at least as early as the discussions between Bohr and Heisenberg in 1927 ([290], pp. 162, 163).

We must be careful here, and we must look beyond 1927 to 1935. In 1927 Bohr was willing to assume that the measurement disturbed the quantum system. The implication was that the source of the measurement problem was this effect that the instrument had on the quantum particle. Bohr understood his error when confronting the issue raised by the paper contributed by Einstein, Boris Podolsky (1896–1966), and Nathan Rosen (1909–1995), the so-called *EPR paper*[6] ([290], p. 178, [102]). The following statement occurs in this short but remarkable paper: "Such a measurement, however, disturbs the particle in a random fashion and thus alters its state." But the authors do not contend that the disturbance is random. They rather point out very clearly that the result is a *reduction of the wave packet*.

Before a quantum mechanical measurement is performed, the state of the quantum system is a sum over all possible states. A measurement returns a single number, which is the value (*eigenvalue*) of the property measured and the state of the quantum system becomes the one corresponding to the eigenvalue obtained in the measurement. Of course, one difficulty here is that the measurement may destroy the quantum system, as when a spot is produced on a film to identify the final state of an electron. But that is not the issue.

The solution to this difficulty is to insist that the measuring instrument is very large compared to the quantum system being measured. Choosing a very small measuring instrument in the hope that this will minimize the effect on the system being measured only introduces the difficulty that both the object of the measurement and the measuring instrument must be treated quantum mechanically.

Bohr's insistence that the measuring instrument be classical and that the results are to be spoken of in classical terms is one solution to this problem. We are only faced with the problem of determining the position of the Heisenberg cut, but that requires only that we do not include quantum effects on the classical side of the cut. Otherwise the cut is arbitrary.

In the autumn of 1927, there were two meetings of the physics community at which Bohr could present his ideas on complementarity. These were the Volta Physics

[6]The paper has the title *Can quantum-mechanical description of physical reality be considered complete?* and is Einstein's fundamental reply to Bohr.

Conference from 11 to 19 September 1927 in Como, Italy, and the Fifth Solvay Conference from 24 to 29 October 1927 in Brussels. The topic at the Solvay Conference was the newly formulated quantum theory. Bohr first introduced his ideas in a very short talk in Como. His ideas were not yet well formulated and the talk was coolly received [125].

At the Solvay Conference, the discussions on complementarity and indeterminism began in earnest over the breakfast table each morning, with Einstein presenting a thought experiment that denied indeterminism. Over the evening meal, Bohr presented a rebuttal showing that the thought experiment of the morning did not lead to a violation of indeterminism ([136], pp. 113, 114).

These discussions, which did not end with the Solvay Conference, are now known as the Bohr–Einstein debates. In a certain sense they concerned the very heart of modern physics. Einstein was convinced that the laws of physics provided an exact description of the universe. According to Heisenberg, Einstein often said, "*Gott würfelt nicht*," (God does not throw dice) during those days in Brussels. And Bohr could only answer with, "*Aber es kann doch nicht unsere Aufgabe sein, Gott vorzuschreiben, wie er die Welt regieren soll.*" (But it is certainly not our responsibility to tell God how to rule the universe.) ([136], p. 115)

In 1930, Heisenberg delivered a series of lectures at the University of Chicago. These are contained in the little book *The Physical Principles of the Quantum Theory*. In the preface, Heisenberg wrote: "But even today the physicist more often has a kind of faith in the correctness of the new principles rather than a clear understanding of them." He then noted that the purpose of the book would be fulfilled if it contributed to the diffusion of the *Kopenhagener Geist der Quantentheorie* (the Copenhagen spirit of the quantum theory), which he claimed should guide the development of modern physics.

What we now mean when we speak of modern physics has moved beyond Heisenberg's understanding when he presented those lectures. The present mathematical understanding of the indeterminacy principle, however, leaves no reason to expect that aspect of the *Kopenhagener Geist* to be set aside.

By insisting that the measurement was the result of an experiment, Bohr tried very hard to remove the concept of measurement from whatever mental picture we may have regarding the form and properties of the quantum particle. This reduced the quantum property to numbers on a screen or a dial. He was very aware, however, that each of us would have our own personal understanding of the quantum particle, or energy parcel.

Many of the experiments we conduct today have moved far beyond those imagined by Bohr and Heisenberg. Whether we still understand experiments in classical terms, as Bohr insisted we must, is a personal question. Historically, however, we cannot deny the importance of the ideas that Heisenberg and Bohr discussed and argued about. Just as Planck feared, the quantum theory has required a new physics that takes us well beyond the realm of classical physics.

If we wish to outline the Copenhagen interpretation, which neither Bohr nor Heisenberg ever did, we should include the following points:

- The quantum theory is a complete mathematical theory with Schrödinger's equation for the wave function as central.
- The wave function itself has no physical meaning, but all physical properties that can be known may be calculated from it.
- Measurements of physical properties are conducted by performing experiments involving quantum systems as objects and classical systems as measuring apparatus.
- Matter has both wave and particle aspects, which are complementary. No experiment can be conducted which measures both wave and particle properties at one and the same time.
- The wave and particle aspects of matter are separated mathematically by Heisenberg's indeterminacy principle.
- All discourse describing quantum mechanics must be conducted in the language of classical mechanics.

By way of explanation, we should note that Schrödinger's equation is taken to be the general form of the equation of motion in quantum mechanics. It may be expressed in operator form and is not restricted to the form of the spatial differential equation we have studied here.

9.5 Summary

We began this chapter with the story of electron spin, a quantum mechanical variable that cannot be understood in the terms used by Lorentz, as we now know.

The interpretation of the wave function became absolutely necessary after Schrödinger's presentation of his new theory in the *Physical Review* paper of 1926. But Born's interpretation must be separated from Schrödinger's contention that the square of the wave function is the electron density. It seems that many of our students still understand the wave function as Schrödinger did. Born's interpretation did not attempt to define what the electron is, but rather where we should expect to find the electron if we choose to make the measurement. We could measure position or momentum, which are well-defined quantities, but we cannot measure the wave function, which is mathematically related to Hamilton's principal function.

We have presented the Copenhagen interpretation which was, as we indicated, the understanding of quantum theory in 1927 or in 1930 when Heisenberg delivered his lectures at the University of Chicago. We have avoided here any pronouncements regarding the validity of any particular understanding of quantum theory. The present author does not subscribe to the belief that interpretation is a personal choice alone. Our understanding of the laws of physics as a community has proven historically to be fluid.

Chapter 10
Connecting the Matrix and Wave Theories

> The symbolic method [...] seems to go more deeply into the nature of things. It enables one to express the physical laws in a neat and concise way.
>
> Paul A. M. Dirac

10.1 Introduction

In our chapter on the Göttingen matrix quantum mechanics, we were dealing with the first quantum theory presented. The approach in this theory was initiated by Heisenberg, essentially alone. The fact that this was based on matrices was noticed by Born. Heisenberg had Pauli's encouragement and Jordan enthusiastically offered his background and talent. The work was finally the product of three minds.

We went through the results carefully. The formulation was based on Hamilton's canonical equations. The quantum condition defined what would have been the canonical transformation in classical theory. Only observable experimental evidence entered the formulation.

At the end of our discussion of the matrix theory, we introduced Dirac, who was working in a quite different research context, and had engaged in a friendly exchange with Heisenberg. We mentioned him because he was then in the process of developing a mathematical approach that would give an insight into the connection between the Göttingen matrix formulation and Schrödinger's formulation.

We have seen that Schrödinger's wave theory was also firmly based on classical mechanics. Indeed, his wave equation was obtained, as we saw, from the Hamilton–Jacobi equation and the link between Fermat's and Hamilton's principles, which had been the subject of part of de Broglie's thesis. However, Schrödinger understood his theory in a way that Heisenberg said was quite impossible. Born's interpretation of the wave function is the one that is now generally accepted. Nevertheless, we still

C. S. Helrich, *The Quantum Theory—Origins and Ideas*, History of Physics,
https://doi.org/10.1007/978-3-030-79268-8_10

have before us two quantum theories that must be reconciled. That is the aim of this chapter.

The mathematical equivalence of the two theories was supposedly first established by Schrödinger [251]. A mathematical proof was provided by von Neumann in his book *Mathematical Foundations of Quantum Mechanics* (see [206], Chap. I). However, von Neumann rejected Dirac's demonstration of the equivalence because of what von Neumann called the fiction of Dirac's δ-function ([206], pp. ix, 25, 27). The Dirac δ-function presently stands on a more rigorous mathematical basis than it did at the time of von Neumann's remark (see, e.g., [257], Chap. 1), so there will be no problem using it here.

We will not attempt to offer a mathematical proof of the equivalence of the two theories. Rather we shall use the fact that the two theories can both be expressed in Dirac's symbolic approach to demonstrate that the actual difference between the two theories lies in whether the time dependence is carried by the operators or the basis vectors.

Our treatment here is based on Dirac's presentation of the equations of motion of a quantum system in Chap. V of the Fourth edition of Dirac's *The Principles of Quantum Mechanics* ([86], pp. 108–116). The quote at the beginning of this chapter is from the first edition of Dirac's book, published in 1930.

10.2 Symbolic Method and Representation

We shall consider here a formulation of the mathematical quantum theory for an undisturbed system in the presence of possible fields of force. If a measurement is performed, the quantum system will be altered in an arbitrary manner as the system state changes from a sum over possible states to a single possible state. Because we must avoid such disturbances, for the sake of mathematical clarity we shall consider only undisturbed systems.

The symbolic quantities of Dirac's approach must be represented by projecting them onto a vector space. This will result in functions we can then treat mathematically. We normally choose the basis vectors to be eigenvectors of the main operator in our problem, which is often the Hamiltonian. Whether these basis vectors are discrete or continuous is determined by the physics of the situation. For example, in the hydrogen atom, the basis vectors (the energy states of the atom) will be discrete, while in the case of a free particle, or of an electron in a crystal lattice, the energy states will be continuous.

In representing our symbolic quantities in terms of the basis vectors of a space, we must make sure that the representation is exact. Mathematically, this is a question of completeness of the space, which requires the set of basis vectors to span the space. This is not a trivial issue and is treated mathematically elsewhere ([61], p. 4; [81], p. 37; [128], p. 294). We shall use Dirac's formulation for the completeness of the set of basis functions, which is more visual and mathematically equivalent.

We identify the set of basis vectors, which span the space, as $|\xi_1\rangle, |\xi_2\rangle, \ldots, |\xi_n\rangle = \{|\xi\rangle\}$, where n may be infinite. The set of basis vectors will also have a dual set $\langle\xi_1|$, $\langle\xi_1|, \ldots, \langle\xi_n| = \{\langle\xi|\}$, which is defined in terms of the scalar product ([86], pp. 18–20). From any set of vectors we can always construct an orthonormal set. Therefore, we assume that our basis vectors are orthonormal without loss of generality. The scalar product for a discrete set of basis vectors is then

$$\langle\xi_i|\,\xi_j\rangle = \delta_{ij} = \begin{cases} 1 \text{ if i } = \text{ j}, \\ 0 \text{ if i } \neq \text{ j}. \end{cases} \tag{10.1}$$

If the basis vectors are continuous, the scalar product of two basis vectors is defined in terms of Dirac's δ-function ([86], pp. 58–61), viz.,

$$\langle\xi'|\,\xi''\rangle = \delta\left(\xi' - \xi''\right). \tag{10.2}$$

The function $\delta\left(\xi' - \xi''\right)$ vanishes for all $\xi' \neq \xi''$ and is infinite at $\xi' = \xi''$. The infinity is such that

$$\int_{-\infty}^{+\infty} f\left(\xi'\right)\delta\left(\xi' - \xi''\right)\mathrm{d}\xi' = f\left(\xi''\right). \tag{10.3}$$

In the fourth edition of his book (1958), Dirac admitted that the δ-function was "not a function according to the usual mathematical definition of a function, which requires a function to have a definite value for each point within its domain," and he called it an improper function. Since then, however, the δ-function has been shown to be representable as a limit of a sequence of test functions (see, e.g., [257], Chap. 1).

To obtain the representation of symbolic functions, Dirac defined a *projection operator*, or *projector*, by

$$\begin{aligned} &\sum_{\xi'} |\xi'\rangle\langle\xi'| \text{ for a discrete basis,} \\ &\int_{\xi'} \mathrm{d}\xi'\, |\xi'\rangle\langle\xi'| \text{ for a continuous basis.} \end{aligned} \tag{10.4}$$

The projection of a vector $|f\rangle$ onto a discrete or continuous basis is

$$\begin{aligned} &\sum_{\xi'} |\xi'\rangle\langle\xi'|\, f\rangle \text{ for a discrete basis,} \\ &\int_{\xi'} \mathrm{d}\xi'\, |\xi'\rangle\langle\xi'|\, f\rangle \text{ for a continuous basis.} \end{aligned} \tag{10.5}$$

Then,

$$\begin{aligned}
\langle\xi'|\, f\rangle &= f_{\xi'} \text{ is the } \xi' \text{ component of } |f\rangle \\
&\quad \text{in the discrete case,} \\
\langle\xi'|\, f\rangle &= f\left(\xi'\right) \text{ is the functional form of } |f\rangle \\
&\quad \text{in the continuous space } |\xi\rangle\,.
\end{aligned} \tag{10.6}$$

If the basis is *complete*, the projection operator is the identity:

$$\begin{aligned}
\sum_{\xi'} |\xi'\rangle\langle\xi'| &= \mathbf{1}, \\
\int_{\xi'} \mathrm{d}\xi'\, |\xi'\rangle\langle\xi'| &= \mathbf{1}.
\end{aligned} \tag{10.7}$$

This is Dirac's more visual formulation of the completeness relation. If the projector is the identity operator, then the space is complete.

The forms of the projectors in (10.7) reveal that, in order to represent an operator in a basis, we must use two projectors. For example, calling the discrete basis vectors $|\mathrm{n}\rangle$ and $|\mathrm{m}\rangle$, the representation of the Hamiltonian $\mathbf{H}$ in a complete discrete basis is

$$\mathbf{H} = \mathbf{1H1} = \sum_{\mathrm{n}}\sum_{\mathrm{m}} |\mathrm{n}\rangle\, \langle\mathrm{n}|\, \mathbf{H}\, |\mathrm{m}\rangle\, \langle\mathrm{m}|\,. \tag{10.8}$$

The representation of the Hamiltonian is then the two-dimensional matrix with elements $\langle\mathrm{n}|\, \mathbf{H}\, |\mathrm{m}\rangle = H_{\mathrm{nm}}$. The representation of the operation of the Hamiltonian on a vector

$$|f\rangle = \sum_{\mathrm{r}} |\mathrm{r}\rangle\, \langle\mathrm{r}|\, f\rangle \tag{10.9}$$

is then

$$\begin{aligned}
\mathbf{H}\,|f\rangle &= \mathbf{1H11}\,|f\rangle \\
&= \sum_{\mathrm{n}}\sum_{\mathrm{m}}\sum_{\mathrm{r}} |\mathrm{n}\rangle\, \langle\mathrm{n}|\, \mathbf{H}\, |\mathrm{m}\rangle\, \langle\mathrm{m}\, |\mathrm{r}\rangle\, \langle\mathrm{r}|\, f\rangle \\
&= \sum_{\mathrm{n}}\sum_{\mathrm{m}} |\mathrm{n}\rangle\, H_{\mathrm{nm}}\, f_{\mathrm{m}},
\end{aligned} \tag{10.10}$$

where we have used (10.1), i.e., $\langle\mathrm{m}\, |\mathrm{r}\rangle = \delta_{\mathrm{mr}}$, and identified $\langle\mathrm{m}|\, f\rangle = f_{\mathrm{m}}$. The operator $\mathbf{H}$ in (10.8) then operates on a vector to produce the vector $\sum_{\mathrm{n}} |\mathrm{n}\rangle \left(\sum_{\mathrm{m}} H_{\mathrm{nm}} f_{\mathrm{m}}\right)$ with elements $\left(\sum_{\mathrm{m}} H_{\mathrm{nm}} f_{\mathrm{m}}\right)$.

Following the same steps in a space with continuous basis vectors, we have

$$
\begin{aligned}
\mathbf{H}\,|f\rangle &= \mathbf{1H11}\,|f\rangle \\
&= \int_{\xi'} \mathrm{d}\xi' \int_{\xi''} \mathrm{d}\xi'' \int_{\xi'''} \mathrm{d}\xi''' \,|\xi'\rangle\langle\xi'|\,\mathbf{H}\,|\xi''\rangle\,\delta\left(\xi''-\xi'''\right)\langle\xi'''|\,f\rangle \\
&= \int_{\xi'} \mathrm{d}\xi' \int_{\xi''} \mathrm{d}\xi'' \,|\xi'\rangle\langle\xi'|\,\mathbf{H}\,|\xi''\rangle\langle\xi''|\,f\rangle\,.
\end{aligned}
\tag{10.11}
$$

Dirac introduced the representation of an operator **Q** in terms of a continuous basis in the form ([86], pp. 69–70)

$$
\langle x'|\,\mathbf{Q}\,|x''\rangle = \mathbf{Q}\left(x'\right)\delta\left(x'-x''\right). \tag{10.12}
$$

Then, using (10.12) and (10.3), Eq. (10.11) becomes

$$
\begin{aligned}
\mathbf{H}\,|f\rangle &= \int_{\xi'} \mathrm{d}\xi' \int_{\xi''} \mathrm{d}\xi'' \,|\xi'\rangle\,\mathbf{H}\left(\xi'\right)\delta\left(\xi'-\xi''\right) f\left(\xi''\right) \\
&= \int_{\xi'} \mathrm{d}\xi' \,|\xi'\rangle\,\mathbf{H}\left(\xi'\right) f\left(\xi'\right),
\end{aligned}
\tag{10.13}
$$

which is a vector in the continuous basis $\left\{|\xi'\rangle\right\}$.

We are considering the motion of a quantum system. All we can know about a dynamical system in classical mechanics is contained in the principal function of Hamilton. And we know that Schrödinger expressed his wave function in terms of the principal function. In the language of Dirac we may call this a "state vector" $|P\rangle$ and represent it as the sum over the possible states available to the system, which in our example are the basis vectors of our space.

Because we are considering dynamical systems, it remains to introduce the time.

10.2.1 Schrödinger Picture

To keep our discussion reasonably simple, we shall assume (with Dirac in this case) that there are two possible system states available to our system: $|A\rangle$ and $|B\rangle$. At the initial time $t = t_0$ the system state, which we shall call $|Rt_0\rangle$ is then

$$
|Rt_0\rangle = c_{\mathrm{A}}\,|At_0\rangle + c_{\mathrm{B}}\,|Bt_0\rangle\,. \tag{10.14}
$$

At a later time t, the system state will be

$$
|Rt\rangle = c_{\mathrm{A}}\,|At\rangle + c_{\mathrm{B}}\,|Bt\rangle\,. \tag{10.15}
$$

We may choose the state vectors $|A\rangle$, $|B\rangle$, and $|R\rangle$ to be normalized to unity and to remain normalized to unity regardless of the time. The constants c_{A} and c_{B} are then unchanged. Each of the state vectors $|A\rangle$ and $|B\rangle$ evolves in time according to the laws of quantum mechanics. We may express this fact in terms of an operator T defined by the requirement that

$$|At\rangle = \mathbf{T}\,|At_0\rangle \text{ and } |Bt\rangle = \mathbf{T}\,|Bt_0\rangle\,.$$

Then,

$$|Rt\rangle = \mathbf{T}\,|Rt_0\rangle = c_{\mathrm{A}}\mathbf{T}\,|At_0\rangle + c_{\mathrm{B}}\mathbf{T}\,|Bt_0\rangle\,. \tag{10.16}$$

To preserve normalization of the state vectors, the operator **T** must be unitary (see Sect. 7.3.4). That is,

$$\mathbf{TT}^{+} = \mathbf{T}^{+}\mathbf{T} = \mathbf{1}.$$

If the state vector $|Pt\rangle$ evolves in time we must be able to define a time derivative of the state vector. This will be

$$\frac{\mathrm{d}}{\mathrm{d}t_0}\,|Pt_0\rangle = \lim_{t\to t_0}\frac{|Pt\rangle - |Pt_0\rangle}{t-t_0} = \lim_{t\to t_0}\frac{\mathbf{T}-\mathbf{1}}{t-t_0}\,|Pt_0\rangle\,, \tag{10.17}$$

calculated at the point t_0. In Appendix E, we obtain the operator equation for the time derivative as (E.21), i.e.,

$$i\hbar\frac{\mathrm{d}}{\mathrm{d}t} = \mathbf{H}, \tag{10.18}$$

where **H** is the Hamiltonian operator and $\hbar = h/2\pi$. Operating with (10.18) on the general state vector $|Pt\rangle$, we have

$$i\hbar\frac{\mathrm{d}}{\mathrm{d}t}\,|Pt\rangle = \mathbf{H}\,|Pt\rangle\,, \tag{10.19}$$

since t_0 is arbitrary ([86], p. 110). Equation (10.18) is Schrödinger's equation for the state vector $|Pt\rangle$.

Reintroducing the operator **T**, (10.18) becomes

$$i\hbar\frac{\mathrm{d}\mathbf{T}}{\mathrm{d}t}\,|Pt_0\rangle = \mathbf{HT}\,|Pt_0\rangle\,. \tag{10.20}$$

Since $|Pt_0\rangle$ is an arbitrary state vector, we may write (10.20) in symbolic operator form as

$$i\hbar\frac{\mathrm{d}\mathbf{T}}{\mathrm{d}t} = \mathbf{HT}. \tag{10.21}$$

To make this look more familiar, we return to (10.19) and choose a representation in terms of continuous coordinates $|\xi\rangle$. Then $|Pt\rangle \to |\psi(t)\rangle$ and (10.19) becomes

$$i\hbar\frac{\partial}{\partial t}\,\langle\xi|\,\psi(t)\rangle = H\,\langle\xi|\,\psi(t)\rangle\,, \tag{10.22}$$

which is Schrödinger's equation in a more familiar form.

In this form of quantum mechanics, the time variable is carried by the time-dependent state vector $|\psi(t)\rangle$. Represented in the coordinates $\{|\xi\rangle\}$, these are the functions $\langle\xi|\ \psi(t)\rangle$, which are the *wave functions* $\psi\,(\xi, t)$ of the Schrödinger theory. Dirac calls this the *Schrödinger picture* of quantum mechanics.

10.2.2 Heisenberg Picture

Since the state vector $|Pt\rangle$ results from the application of the operator **T** to the initial state vector $|Pt_0\rangle$, we may identify the original state vector at time t_0 as

$$|Pt_0\rangle = \mathbf{T}^{-1}\,|Pt\rangle\,. \tag{10.23}$$

If we transform all the vectors in our system back in time in this way, we must also transform the operators as well. Hence, we must work out how to transform operators in time.

We assume that at time t we have the state vector $|St\rangle$ obtained from the state vector $|Pt\rangle$ by action of the operator **Q** as

$$|St\rangle = \mathbf{Q}\,|Pt\rangle\,. \tag{10.24}$$

In (10.24), both state vectors $|St\rangle$ and $|Pt\rangle$ are time dependent and the operator **Q** is independent of the time. We now transform this result back to the initial condition:

$$\begin{aligned}|St_0\rangle &= \mathbf{T}^{-1}\,|St\rangle = \mathbf{T}^{-1}\mathbf{Q}\,|Pt\rangle \\ &= \mathbf{T}^{-1}\mathbf{Q}\mathbf{T}\mathbf{T}^{-1}\,|Pt\rangle \\ &= \mathbf{T}^{-1}\mathbf{Q}\mathbf{T}\,|Pt_0\rangle\,,\end{aligned} \tag{10.25}$$

where we have used $\mathbf{T}\mathbf{T}^{-1} = \mathbf{1}$.

This is the form our operation in (10.24) will take if the time variable is carried by the operators rather than by the state vectors. Specifically, the time-independent operator **Q** becomes the time-dependent operator $\mathbf{Q}_\mathrm{t}$ by the operation

$$\mathbf{Q} \to \mathbf{Q}_\mathrm{t} = \mathbf{T}^{-1}\mathbf{Q}\mathbf{T}. \tag{10.26}$$

Operation of **Q** upon the state vector $|Pt\rangle$ will then become

$$\mathbf{Q}\,|Pt\rangle \to \mathbf{Q}_\mathrm{t}\,|Pt_0\rangle = \mathbf{T}^{-1}\mathbf{Q}\mathbf{T}\,|Pt_0\rangle\,, \tag{10.27}$$

if the time is carried by the operators. We explicitly indicate the time dependence of the operator $\mathbf{Q}_\mathrm{t}$ by the subscript **t**.

We now have the operator equation

$$\mathbf{T}\mathbf{Q}_t = \mathbf{Q}\mathbf{T}. \tag{10.28}$$

The time derivative of (10.28) is

$$\frac{d\mathbf{T}}{dt}\mathbf{Q}_t + \mathbf{T}\frac{d\mathbf{Q}_t}{dt} = \mathbf{Q}\frac{d\mathbf{T}}{dt}, \tag{10.29}$$

since $\mathbf{Q}$ is time independent. From (10.21), Eq. (10.29) becomes

$$\mathbf{H}\mathbf{T}\mathbf{Q}_t + i\hbar\mathbf{T}\frac{d\mathbf{Q}_t}{dt} = \mathbf{Q}\mathbf{H}\mathbf{T}. \tag{10.30}$$

Rearranging (10.30) and applying $\mathbf{T}^{-1}$ to both sides, we have

$$\begin{aligned} i\hbar\frac{d\mathbf{Q}_t}{dt} &= \left(\mathbf{T}^{-1}\mathbf{Q}\mathbf{T}\right)\mathbf{T}^{-1}\mathbf{H}\mathbf{T} - \mathbf{T}^{-1}\mathbf{H}\mathbf{T}\left(\mathbf{Q}_t\right) \\ &= \mathbf{Q}_t\mathbf{H}_t - \mathbf{H}_t\mathbf{Q}_t, \end{aligned} \tag{10.31}$$

where $\mathbf{H}_t = \mathbf{T}^{-1}\mathbf{H}\mathbf{T}$. We now recall the Eq. (7.72) from Born and Jordan's paper on the matrix theory. In the present notation, this is

$$i\hbar\dot{\mathbf{f}} = (\mathbf{f}\mathbf{H} - \mathbf{H}\mathbf{f}). \tag{10.32}$$

We thus recover one of the main dynamical equations from matrix quantum mechanics if we choose to let the operators carry the time dependence.

If we use Dirac's classical analog for the Poisson bracket (PB), viz., (7.100), we see that (10.31) is the quantum mechanical equivalent of the classical equation for a function Q:

$$\frac{dQ}{dt} = \{Q, H\},$$

where $\{\cdots\}$ is a PB.

We have discovered that the matrices of the Göttingen formulation of quantum mechanics are the matrix representation of the symbolic operators of the Dirac formulation. The matrix quantities carry the time dependence, as the symbolic operators $\mathbf{Q}_t$ do here. This is what Dirac calls the *Heisenberg picture*. Dirac thereby reduced the difference between the wave mechanical formulation of Schrödinger and the matrix formulation of Born, Heisenberg, and Jordan to a choice of where to introduce the time variable.

10.3 Summary

In this brief chapter we have established the fundamental difference between the matrix and wave formulations of quantum theory. This was not intended as a mathematical proof of equivalence. We have only presented an abbreviated form of Dirac's discussion of the relationship between what he called the Schrödinger and Heisenberg pictures of the quantum theory.

This has required some introduction to the language of Dirac's symbolic approach, which has now become the standard language of the quantum theory in any beginning course. So this introduction is not out of place, even if our main objective is the origins of the quantum theory.

Chapter 11
Epilogue

I think I can safely say that nobody understands quantum mechanics.
Richard P. Feynman

The above quote is from Richard Feynman's (1918–1988) *Messenger Lectures* in 1964 at MIT. Feynman came to this conclusion after referring to a newspaper article which stated that there were only twelve people in the world who understood Einstein's relativity. He pointed out that, shortly after Einstein's theory appeared in print in 1905, it was understood in some way by many more people than twelve. His qualifier "in some way" indicates that there may not have been an accepted sense in which that understanding could be be checked. Nevertheless, we might well feel that Feynman's claim that nobody understands quantum mechanics is valid after the study we have just undertaken.

Our study has been about what happened to the way in which we understand, or believe we understand, the universe after encountering the quantum. I have tried to provide the history without further comment in order to leave the final judgement to the reader. This history is also longer than traditional introductions would suggest, and it is nuanced by the personal sense of each of the participants regarding the meaning of physics, physical law, and experiment. The reader has at least had the opportunity to hear others with deep questions and to decide for themself what understanding quantum mechanics may actually entail.

My own understanding and experience of this period of history have been altered by writing this book. We have indeed encountered giants as we walked this road. But in this drama, all the actors, participating at all levels, were important. At each step along the road, or the mountain path, no single student or assistant knew what the outcome would be. The actors really had no script. In fact, their problem was to develop the script. If the student was fortunate, their mentor was a teacher like Sommerfeld or Ehrenfest, who could help them see the importance of that Alpine path. And that is exactly the issue that faces all our students even today. They are trying to find their script.

C. S. Helrich, *The Quantum Theory—Origins and Ideas*, History of Physics,
https://doi.org/10.1007/978-3-030-79268-8_11

I began this study with the desire to provide students with a deeper understanding of how physical science works. That was my primary motivation as I sought to link up that wonderful but disconnected set of experiments to which we expose our students at the beginning of every course on quantum mechanics. How many of us ever told our students why blackbody radiation was so interesting and that it occupied a central place in one of the world's most important laboratories? Or why exactly the head of the Cavendish Laboratory considered that the identity of the cathode ray was the central issue in English physics? For that matter, how many of our students know the importance of scientific advances to the survival of civilization? Connecting these steps into a consistent story led me to the first ideas of the Presocratic philosophers and through the thoughts of Davy and Faraday to the laboratories in Berlin and Cambridge.

We cannot separate the history, if we may call it that, from the mathematics. This has been the case since the earliest ventures into philosophy. Failure to recognize this, it seems to me, has resulted in our disparaging remarks about, for example, Thomson's model of the atom. And we may fail to notice the beauty of Einstein's development of the photon concept if we attempt to simplify it to an explanation of the photoelectric effect. We have often tried to simplify de Broglie's contribution to a single result, the one relating the wavelength to the particle momentum. And Schrödinger's use of mathematics, and the conclusions he attempted to draw, form a topic of study on their own. I hope that my outline here will at least open some doors to future students. That was my main purpose.

The last chapter had a dual purpose. After presenting two apparently different quantum theories, something must be said about reconciling them. Fortunately, Dirac's symbolic approach provides a clear mathematical platform from which to accomplish this task. We elected here to use the basic presentation Dirac provided in a chapter of his book. It is no longer remarkable to note that there is no actual difference between the matrix and the wave mechanical theories. Dirac refers to these as the Schrödinger and the Heisenberg pictures, rather than theories. They differ only in the choice of where we introduce the time variable. This discovery unifies these supposedly different theories, and it may make the transition to the final state of the theory more understandable for students.

Of course, such a story has no end, and maybe we hope it has no end. We must, however, end our contribution someplace. Perhaps at this point the reader will at least be in a position to respond to Feynman's claim.

Appendix A
Hamilton and Fermat

A central step in Schrödinger's development was his contention that

$$J = \int \mathrm{d}\mathbf{r} \left[\left(\frac{\partial \psi}{\partial x} \right)^2 + \left(\frac{\partial \psi}{\partial y} \right)^2 + \left(\frac{\partial \psi}{\partial z} \right)^2 - \frac{2m}{K^2} \left(\mathcal{E} - V\left(r \right) \right) \psi^2 \right]$$

has an extremum for all real-valued and twice differentiable functions ψ. This provided a way round the non-linear equation

$$\left(\frac{\partial \psi}{\partial x} \right)^2 + \left(\frac{\partial \psi}{\partial y} \right)^2 + \left(\frac{\partial \psi}{\partial z} \right)^2 - \frac{2m}{K^2} \left(\mathcal{E} - V \right) \psi^2 = 0$$

to the rather familiar and soluble equation

$$\left(\frac{\partial^2}{\partial x^2} \psi \right) + \left(\frac{\partial^2}{\partial y^2} \psi \right) + \left(\frac{\partial^2}{\partial z^2} \psi \right) + \frac{2m}{K^2} \left(\mathcal{E} - V\left(r \right) \right) \psi = 0.$$

We will show here that this step is justified by the equivalence of Fermat's and Jacobi's principles. Jacobi's principle is equivalent to Hamilton's principle and de Broglie cited the origins of Hamilton's ideas in Maupertuis. In de Broglie's words:

> Fermat's principle applied to a phase wave is equivalent to Maupertuis' principle applied to a particle in motion. The possible trajectories of the particle are identical to the rays of the phase wave.

We shall now develop this idea, although our approach will not follow de Broglie's exactly.

C. S. Helrich, *The Quantum Theory—Origins and Ideas*, History of Physics,
https://doi.org/10.1007/978-3-030-79268-8

A.1 W and ϕ Surfaces

We begin with the Hamilton–Jacobi equation

$$\frac{\partial S}{\partial t} + \mathcal{H}\left(\frac{\partial S}{\partial q_1}, \ldots, \frac{\partial S}{\partial q_3}, q_1, \ldots, q_3\right) = 0, \tag{A.1}$$

where $p_i = \partial S\,(q, t)\,/\partial q_i$ is the ith component of the momentum. For simplicity, we shall confine ourselves to Cartesian space. There will then be only three coordinates $q = (q_1, q_2, q_3)$.

We now carry out a general separation of (A.1) by writing $S\,(q, t)$ as

$$S\,(q, t) = S_t(t) + W(q). \tag{A.2}$$

Using (A.2), Eq. (A.1) becomes

$$\frac{\mathrm{d}S_t(t)}{\mathrm{d}t} = -\mathcal{H}\left(\frac{\partial W}{\partial q_1}, \ldots, \frac{\partial W}{\partial q_3}, q_1, \ldots, q_3\right). \tag{A.3}$$

The left-hand side of (A.3) depends only on the time and the right-hand side depends only on the spatial coordinates. Since the time and the spatial coordinates are independent variables, each side of (A.3) must be equal to a constant. If we consider conservative systems, the Hamiltonian $\mathcal{H}$ is numerically equal to the energy $\mathcal{E}$. Hence, $S_t(t) = -\mathcal{E}t$ and (A.2) becomes

$$S\,(q, t) = -\mathcal{E}t + W(q). \tag{A.4}$$

The function $W(q)$ then satisfies

$$\mathcal{H}\left(\frac{\partial W(q)}{\partial q_1}, \ldots, q_1, \ldots\right) = \mathcal{E}. \tag{A.5}$$

With the Hamiltonian for a conservative system, viz.,

$$\mathcal{H} = \frac{1}{2m}\sum_j p_j^2 + V\,(q)\,. \tag{A.6}$$

Equation (A.5) becomes

$$\sum_j \left(\frac{\partial W(q)}{\partial q_j}\right)^2 = 2m\,[\mathcal{E} - V\,(q)]\,. \tag{A.7}$$

This partial differential equation specifies a set of surfaces $W\,(q)$ in space.

The particle momenta $p_i = \partial W(q)/\partial q_i$ are components of the gradient of the function $W(q)$. Then Eq. (A.7) is

$$|\text{grad } W| = \sqrt{2m\,[\mathcal{E}-V(q)]} = p_W, \tag{A.8}$$

where p_W is the magnitude of the particle momentum.

Since grad W is the directional derivative perpendicular to the surface W, we may introduce q_W as the coordinate normal to the surface W and write

$$\text{grad } W = \frac{\partial W}{\partial q_W}\hat{n}_W = p_W \hat{n}_W, \tag{A.9}$$

where $\hat{n}_W$ is the unit vector normal to the surface W of constant momentum.

There are an infinite number of possible particle trajectories each crossing a particular $W(q)$ surface at a different point. For each trajectory,

$$\frac{\partial W}{\partial q_W} = p_W = \sqrt{2m\,[\mathcal{E}-V(q)]}, \tag{A.10}$$

where the point is specified by q.

We may assign an arbitrary numerical value to a particular W-surface, and then the values of the other W-surfaces follow. The difference in values between any two W-surfaces is, however, not arbitrary and follows from

$$\begin{aligned}\Delta W_{12} &= \int_{W_1}^{W_2} \frac{\partial W}{\partial q_W}\hat{n}_W \cdot \mathbf{dq} \\ &= \int_{W_1}^{W_2} \sqrt{2m\,[\mathcal{E}-V(q)]}\mathrm{d}C,\end{aligned} \tag{A.11}$$

where $\mathrm{d}C = \hat{n}_W \cdot \mathbf{dq}$ is the differential distance along the particle trajectory in configuration space. The integration in (A.11) is along the particle trajectory.

The integral (A.11) is called *Jacobi's integral. Jacobi's principle* requires this integral to be a *minimum*. Jacobi's principle is logically equivalent to Hamilton's principle ([165], pp. 132–135, 273).

We now turn to Christiaan Huygens' (1629–1695) wave principle for optics in differential form. This equation for the phase wave[1] surfaces is ([165], p. 270)

$$\sum_j \left(\frac{\partial \phi}{\partial q_j}\right)^2 = \frac{n^2(q)}{c^2}, \tag{A.12}$$

[1] A finite optical (light) wave will be a sum over a group of these phase waves.

where $n(q)$ is the index of refraction of the medium through which the light is passing and c is the speed of light in vacuum. The index of refraction may be spatially dependent, which we indicate by including a dependence on the coordinate q in n.

Equation (A.12) was derived by Hamilton ([165], p. 270; [213], p. 235). The curved surfaces $\phi(x, y, z)$ are the wavefronts of Huygens and the velocity of light in the medium to which (A.12) applies is

$$v_\phi(q) = \frac{c}{n(q)}. \tag{A.13}$$

As for the W-surfaces, we may write

$$\text{grad}\,\phi = \frac{\partial\phi}{\partial q_\phi}\hat{n}_\phi = \frac{1}{v_\phi(q)}\hat{n}_\phi, \tag{A.14}$$

where q_ϕ and $\hat{n}_\phi$ are the coordinate and the unit vector along the normal to the surface ϕ at the point q. We have used the subscript ϕ to indicate that the light is carried by the phase waves described by Huygens' equation.

Analogously to the surfaces W, the surfaces ϕ determine the speed of light on each surface. As the gradient of ϕ increases, however, the velocity v_ϕ decreases. That is, an increasing index of refraction decreases the speed of light passing through the medium.

If $\mathrm{d}t_\phi$ denotes the time for light with speed $v_\phi(x, y, z)$ to travel between the surfaces ϕ and $\phi+\mathrm{d}\phi$, then the distance $\mathrm{d}q_\phi$ between two surfaces ϕ and $\phi+\mathrm{d}\phi$ is

$$\mathrm{d}q_\phi = v_\phi(x, y, z)\,\mathrm{d}t_\phi.$$

The time $\Delta\tau_{12} = \tau_2 - \tau_1$ that light requires to pass between the surfaces ϕ_1 and ϕ_2 is then

$$\Delta\tau_{12} = \int_{\tau_1}^{\tau_2} \mathrm{d}t_\phi = \int_{q_1}^{q_2} \frac{\mathrm{d}q_\phi}{v_\phi} = \int_{q_1}^{q_2} \frac{n(q)}{c}\mathrm{d}q_\phi. \tag{A.15}$$

The differential $\mathrm{d}q_\phi$ is along the path followed by the ray of light. Equation (A.15) gives the time taken for light to follow the path determined by the index of refraction of the medium. *Fermat's principle* requires the path taken by the light beam to be the one that minimizes the time $\Delta\tau_{12}$.

We then have two statements: (1) *Jacobi's principle* in analytical mechanics, which requires ΔW_{12} as defined in (A.11) to be a minimum for the correct particle trajectory in a potential $V(q)$, and (2) *Fermat's principle* in optics, which requires $\Delta\tau_{12}$ as defined in (A.15) to be a minimum for the correct path of a light beam in a medium with index of refraction $n(q)$. These are analogous if the integrands in (A.11) and (A.15) differ by only a constant factor, that is, if

$$\frac{n(q)}{c} = \alpha\sqrt{2m\,(\mathcal{E} - V\,(q))}, \tag{A.16}$$

where α is that constant factor.

With (A.13) and (A.10), Eq. (A.16) becomes

$$\frac{1}{v_\phi(q)} = \alpha p_{\mathrm{W}}(q). \tag{A.17}$$

Hence, the phase velocity of the (light) waves obtained from Huygens' equation and the velocity of the particle obtained from the Hamilton–Jacobi equation are inversely proportional to one another.

The left-hand side of (A.17) has physical dimensions $[time/length]$, and the right-hand side has dimensions $[mass \cdot length/time]\,[\alpha]$. Therefore the dimensions of α must be $[time^2/mass \cdot length^2]$. We can form the constant α from the frequency of the optical wave ν_ϕ and Planck's constant h as

$$\alpha = \frac{B}{h\nu_\phi}, \tag{A.18}$$

where B is an arbitrary (dimensionless) number. Using (A.18) and $v_\phi = \nu_\phi \lambda_\phi$, where ν_ϕ and λ_ϕ are the frequency and wavelength of the light wave, (A.17) becomes

$$\lambda_\phi = \frac{h}{Bp_{\mathrm{W}}}. \tag{A.19}$$

Then, with $B = 1$, this is the relationship de Broglie found for the moving particle. We have thus found that de Broglie's idea connects Huygens' and Jacobi's (Hamilton's) principles for geometrical optics and the motion of material particles based on the equivalence of Fermat's and Jacobi's (Hamilton's) principles. This point was made by de Broglie in his thesis. De Broglie, however, cited Maupertuis as the originator of the mechanical principle, rather than Hamilton.

A.2 Variational Principle

Jacobi's principle requires the integral of $\sqrt{2m\,(\mathcal{E} - V(q))}$ over all space to vanish. That is,

$$\delta \int_{\mathrm{W}_1}^{\mathrm{W}_2} \sqrt{2m\,(\mathcal{E} - V(q))}\mathrm{d}q_{\mathrm{W}} = 0\,, \tag{A.20}$$

along all paths between the surfaces W_1 and W_2, which are perpendicular to the surfaces W.

We may construct a small cylinder around the infinitesimal path $\mathrm{d}q_{\mathrm{W}}$ and then integrate over the volume of this cylinder between the surfaces W_1 and W_2. The integral over the entire volume between the two surfaces is thus obtained by integrating over all cylinders around each dq_W and then over all paths dq_W. We then simply change the integration in (A.20) to a volume integral between the surfaces as

$$\delta \int \mathbf{dr}\,[2m\,(\mathcal{E} - V\,(r))] = 0, \tag{A.21}$$

where we have dropped the square root as unnecessary and introduced the spatial volume differential $\mathbf{dr}$, since the integration is now over spatial variables.

Before introducing $S = K \ln \psi$, Schrödinger's partial differential equation was

$$\left(\frac{\partial W\,(r)}{\partial x}\right)^2 + \left(\frac{\partial W\,(r)}{\partial y}\right)^2 + \left(\frac{\partial W\,(r)}{\partial z}\right)^2 - 2m\,(\mathcal{E} - V) = 0 \tag{A.22}$$

and the integral J was

$$J = \int \mathbf{dr}\left[\left(\frac{\partial W\,(r)}{\partial x}\right)^2 + \left(\frac{\partial W\,(r)}{\partial y}\right)^2 + \left(\frac{\partial W\,(r)}{\partial z}\right)^2 - 2m\,(\mathcal{E} - V\,(r))\right]. \tag{A.23}$$

The Hamilton–Jacobi equation guarantees that $J = 0$, but this value of J is an extremum only if $\delta J = 0$.

Since $J = 0$, and using (A.21), we can write the variation of (A.23) as

$$\delta \int \mathbf{dr}\left[\left(\frac{\partial W\,(r)}{\partial x}\right)^2 + \left(\frac{\partial W\,(r)}{\partial y}\right)^2 + \left(\frac{\partial W\,(r)}{\partial z}\right)^2\right] = \delta \int \mathbf{dr}2m\,(\mathcal{E} - V\,(r)) = 0. \tag{A.24}$$

Therefore,

$$\delta \int \mathbf{dr}\left[\left(\frac{\partial W\,(r)}{\partial x}\right)^2 + \left(\frac{\partial W\,(r)}{\partial y}\right)^2 + \left(\frac{\partial W\,(r)}{\partial z}\right)^2 - 2m\,(\mathcal{E} - V\,(r))\right] = 0, \tag{A.25}$$

which is the variational principle Schrödinger introduced at the bottom of the first page of his first communication. Imposing this requirement guarantees that the particle trajectory will follow the optical path of the phase waves. This is, of course, central to the theory Schrödinger was developing.

Appendix B
Schrödinger's Variation

Here we carry out the details of the variation δJ of the functional J defined by

$$J = \int \mathbf{dr} \left[\left(\frac{\partial \psi}{\partial x} \right)^2 + \left(\frac{\partial \psi}{\partial y} \right)^2 + \left(\frac{\partial \psi}{\partial z} \right)^2 - \frac{2m}{K^2} \left(\mathcal{E} - V(r) \right) \psi^2 \right]. \tag{B.1}$$

The integral is over the entire spatial volume containing the system (all space). The variation of this integral functional will be subject to the requirement that the variation $\delta\psi$ in ψ is arbitrary within the volume considered and vanishes on the boundary of the volume.

The variation of J is

$$\begin{aligned} \delta J = \int \mathbf{dr} \Bigg[& \left(\frac{\partial (\psi + \delta\psi)}{\partial x} \right)^2 + \left(\frac{\partial (\psi + \delta\psi)}{\partial y} \right)^2 \\ & + \left(\frac{\partial (\psi + \delta\psi)}{\partial z} \right)^2 - \frac{2m}{K^2} \left(\mathcal{E} - V(r) \right) (\psi + \delta\psi)^2 \Bigg] - J. \end{aligned} \tag{B.2}$$

Neglecting terms quadratic in $\delta\psi$, Eq. (B.2) becomes

$$\begin{aligned} \delta J = \int \mathbf{dr} \Bigg[& \left(\frac{\partial \psi}{\partial x} \right)^2 + 2 \frac{\partial \psi}{\partial x} \frac{\partial \delta\psi}{\partial x} \\ & + \left(\frac{\partial \psi}{\partial y} \right)^2 + 2 \frac{\partial \psi}{\partial y} \frac{\partial \delta\psi}{\partial y} + \left(\frac{\partial \psi}{\partial z} \right)^2 + 2 \frac{\partial \psi}{\partial z} \frac{\partial \delta\psi}{\partial z} \\ & - \frac{2m}{K^2} \left(\mathcal{E} - V(r) \right) \left(\psi^2 + 2\varphi\delta\psi \right) \Bigg] - J. \end{aligned} \tag{B.3}$$

Subtracting the terms present in the integral J, Eq. (B.3) is

C. S. Helrich, *The Quantum Theory—Origins and Ideas*, History of Physics,
https://doi.org/10.1007/978-3-030-79268-8

$$\delta J = 2\int \mathbf{dr}\left[\frac{\partial\psi}{\partial x}\frac{\partial\delta\psi}{\partial x} + \frac{\partial\psi}{\partial y}\frac{\partial\delta\psi}{\partial y} + \frac{\partial\psi}{\partial z}\frac{\partial\delta\psi}{\partial z} - \frac{2m}{K^2}\left(\mathcal{E} - V(r)\right)(\psi\delta\psi)\right]. \tag{B.4}$$

We now carry out a partial integration, noting that

$$\frac{\partial\psi}{\partial x}\frac{\partial\delta\psi}{\partial x} = \frac{\partial}{\partial x}\frac{\partial\psi}{\partial x}\delta\psi - \frac{\partial^2\psi}{\partial x^2}\delta\psi.$$

The integral in (B.4) then becomes

$$\delta J = 2\int \mathbf{dr}\left(\frac{\partial}{\partial x}\frac{\partial\psi}{\partial x}\delta\psi + \frac{\partial}{\partial y}\frac{\partial\psi}{\partial y}\delta\psi + \frac{\partial}{\partial z}\frac{\partial\psi}{\partial z}\delta\psi\right) + 2\int \mathbf{dr}\left[-\frac{\partial^2\psi}{\partial x^2}\delta\psi - \frac{\partial^2\psi}{\partial y^2}\delta\psi - \frac{\partial^2\psi}{\partial z^2}\delta\psi - \frac{2m}{K^2}\left(\mathcal{E} - V(r)\right)(\psi\delta\psi)\right]. \tag{B.5}$$

The first integral in (B.5) is

$$\int \mathbf{dr}\left(\frac{\partial}{\partial x}\frac{\partial\psi}{\partial x}\delta\psi + \frac{\partial}{\partial y}\frac{\partial\psi}{\partial y}\delta\psi + \frac{\partial}{\partial z}\frac{\partial\psi}{\partial z}\delta\psi\right) = \int \mathbf{dr}\,\mathrm{div}\,\mathrm{grad}\,\psi\delta\psi. \tag{B.6}$$

Applying Gauss' theorem to (B.6), we have

$$\int \mathbf{dr}\,\mathrm{div}\,\mathrm{grad}\,\psi\delta\psi = \oint_{\mathrm{S}} (\mathrm{grad}\,\psi\delta\psi)\cdot \mathbf{dS} = 0, \tag{B.7}$$

because $\delta\psi$ vanishes on the boundary. The variation in (B.5) is then

$$\delta J = 2\int \mathbf{dr}\left[-\frac{\partial^2\psi}{\partial x^2} - \frac{\partial^2\psi}{\partial y^2} - \frac{\partial^2\psi}{\partial z^2} - \frac{2m}{K^2}\left(\mathcal{E} - V(r)\right)\psi\right]\delta\psi. \tag{B.8}$$

Because the variation $\delta\psi$ is arbitrary within the volume, the requirement $\delta J = 0$ is satisfied if and only if

$$\frac{\partial^2\psi}{\partial x^2} + \frac{\partial^2\psi}{\partial y^2} + \frac{\partial^2\psi}{\partial z^2} + \frac{2m}{K^2}\left(\mathcal{E} - V(r)\right)\psi = 0. \tag{B.9}$$

This is the equation Schrödinger obtained (see (8.10)).

In §7 of the *Physical Review* (1926) paper, Schrödinger referred to the results of a general variation in which he considered a conservative system with what he called a completely arbitrary Hamiltonian

$$\mathcal{H}_{\mathrm{gen}} = \frac{1}{2}\sum_{\mathrm{j,k=1}}^{\mathrm{N}} a_{\mathrm{jk}} p_{\mathrm{j}} p_{\mathrm{k}} + V. \tag{B.10}$$

The integral (8.38) is then

$$I_{\mathrm{gen}} = \int \mathbf{dq}\left[\frac{(h/2\pi)^2}{2m}\sum_{\mathrm{j,k}}^{\mathrm{N}} a_{\mathrm{jk}}\frac{\partial\psi}{\partial q_{\mathrm{j}}}\frac{\partial\psi}{\partial q_{\mathrm{k}}} - (\mathcal{E} - V(r))\,\psi^2\right], \tag{B.11}$$

where $\mathbf{dq} = \mathrm{d}q_1 \cdots \mathrm{d}q_{\mathrm{N}}$.[2]

The variation is defined by

$$\delta I_{\mathrm{gen}} = \int \mathbf{dq}\left[\frac{\hbar^2}{2m}\sum_{\mathrm{j,k}}^{\mathrm{N}} a_{\mathrm{jk}}\frac{\partial(\psi+\delta\psi)}{\partial q_{\mathrm{j}}}\frac{\partial(\psi+\delta\psi)}{\partial q_{\mathrm{k}}}\right.$$
$$\left. - (\mathcal{E} - V(r))(\psi+\delta\psi)^2\right] - I. \tag{B.12}$$

Expanding the partial derivative terms to first order in the variation $\delta\psi$, we have

$$\frac{\partial(\psi+\delta\psi)}{\partial q_{\mathrm{j}}}\frac{\partial(\psi+\delta\psi)}{\partial q_{\mathrm{k}}}$$
$$\approx \frac{\partial\psi}{\partial q_{\mathrm{j}}}\frac{\partial\psi}{\partial q_{\mathrm{k}}} + \frac{\partial\psi}{\partial q_{\mathrm{j}}}\frac{\partial\delta\psi}{\partial q_{\mathrm{k}}} + \frac{\partial\psi}{\partial q_{\mathrm{k}}}\frac{\partial\delta\psi}{\partial q_{\mathrm{j}}} \tag{B.13}$$

and

$$(\psi+\delta\psi)^2 \approx \psi^2 + 2\psi\delta\psi. \tag{B.14}$$

Then (B.12) becomes

$$\delta I_{\mathrm{gen}} = \int \mathbf{dq}\left[\frac{\hbar^2}{2m}\sum_{\mathrm{j,k}}^{\mathrm{N}} a_{\mathrm{jk}}\left(\frac{\partial\psi}{\partial q_{\mathrm{j}}}\frac{\partial\delta\psi}{\partial q_{\mathrm{k}}}\right.\right.$$
$$\left.\left. + \frac{\partial\psi}{\partial q_{\mathrm{k}}}\frac{\partial\delta\psi}{\partial q_{\mathrm{j}}}\right) - (\mathcal{E} - V(r))(2\psi\delta\psi)\right]. \tag{B.15}$$

---

[2] If we are considering a transformation of coordinates, there must be a Jacobian for the transformation here ([138], pp. 378–379). We are beginning our development from the generalized coordinates $\mathbf{q}$, however, rather than transforming coordinates.

To integrate the terms

$$\frac{\partial \psi}{\partial q_j}\frac{\partial \delta\psi}{\partial q_k} + \frac{\partial \psi}{\partial q_k}\frac{\partial \delta\psi}{\partial q_j},$$

we proceed as we would for a partial integration:

$$\begin{aligned}\frac{\partial}{\partial q_j}\frac{\partial \psi\delta\psi}{\partial q_k} &= \frac{\partial}{\partial q_j}\left(\psi\frac{\partial \delta\psi}{\partial q_k} + \delta\psi\frac{\partial \psi}{\partial q_k}\right)\\ &= \left(\frac{\partial \psi}{\partial q_j}\frac{\partial \delta\psi}{\partial q_k} + \psi\frac{\partial^2 \delta\psi}{\partial q_j \partial q_k}\right.\\ &\quad \left. + \frac{\partial \delta\psi}{\partial q_j}\frac{\partial \psi}{\partial q_k} + \delta\psi\frac{\partial \psi}{\partial q_j \partial q_k}\right). \qquad (B.16)\end{aligned}$$

Then

$$\begin{aligned}\frac{\partial \psi}{\partial q_j}\frac{\partial \delta\psi}{\partial q_k} &= \frac{\partial}{\partial q_j}\frac{\partial \psi\delta\psi}{\partial q_k} - \psi\frac{\partial^2 \delta\psi}{\partial q_j \partial q_k}\\ &\quad - \frac{\partial \delta\psi}{\partial q_j}\frac{\partial \psi}{\partial q_k} - \delta\psi\frac{\partial \psi}{\partial q_j \partial q_k} \qquad (B.17)\end{aligned}$$

and

$$\begin{aligned}\frac{\partial \psi}{\partial q_k}\frac{\partial \delta\psi}{\partial q_j} &= \frac{\partial}{\partial q_k}\frac{\partial \psi\delta\psi}{\partial q_j} - \psi\frac{\partial^2 \delta\psi}{\partial q_j \partial q_k}\\ &\quad - \frac{\partial \delta\psi}{\partial q_k}\frac{\partial \psi}{\partial q_j} - \delta\psi\frac{\partial \psi}{\partial q_j \partial q_k}. \qquad (B.18)\end{aligned}$$

Adding (B.17) and (B.18), we obtain

$$\begin{aligned}&\left(\frac{\partial \psi}{\partial q_j}\frac{\partial \delta\psi}{\partial q_k} + \frac{\partial \psi}{\partial q_k}\frac{\partial \delta\psi}{\partial q_j}\right)\\ &= 2\frac{\partial^2 \psi\delta\psi}{\partial q_j \partial q_k} - 2\psi\frac{\partial^2 \delta\psi}{\partial q_j \partial q_k}\\ &\quad - 2\frac{\partial \delta\psi}{\partial q_j}\frac{\partial \psi}{\partial q_k} - 2\delta\psi\frac{\partial^2 \psi}{\partial q_j \partial q_k}, \qquad (B.19)\end{aligned}$$

since the order of partial differentiation is immaterial.

We now neglect the curvature $\partial^2\delta\psi/\partial q_j \partial q_k$ and the gradient $\partial\delta\psi/\partial q_j$ of $\delta\psi$, considering them to be small, since $\delta\psi$ is infinitesimal. Then (B.19) becomes

$$\left(\frac{\partial\psi}{\partial q_{\mathrm{j}}}\frac{\partial\delta\psi}{\partial q_{\mathrm{k}}}+\frac{\partial\psi}{\partial q_{\mathrm{k}}}\frac{\partial\delta\psi}{\partial q_{\mathrm{j}}}\right)$$
$$=2\frac{\partial^2\psi\delta\psi}{\partial q_{\mathrm{j}}\partial q_{\mathrm{k}}}-2\delta\psi\frac{\partial^2\psi}{\partial q_{\mathrm{j}}\partial q_{\mathrm{k}}}. \tag{B.20}$$

If $j \neq k$, the integral over the first term on the right, viz.,

$$\int \mathbf{dq}\sum_{\mathrm{k}}^{\mathrm{N}}\frac{\partial}{\partial q_{\mathrm{j}}}\frac{\partial\psi\delta\psi}{\partial q_{\mathrm{k}}},$$

contains the integral

$$\int_{-\infty}^{+\infty} dq_{\mathrm{k}}\frac{\partial\psi\delta\psi}{\partial q_{\mathrm{k}}} = \psi\delta\psi]_{-\infty}^{+\infty} = 0$$

as a factor, and this vanishes because $\delta\psi = 0$ on the boundary. If $j = k$, the integral over the terms $\psi\partial^2\delta\psi/\partial q_{\mathrm{j}}\partial q_{\mathrm{k}}$ is

$$\int \mathbf{dq}\sum_{\mathrm{k}}^{\mathrm{N}}\frac{\partial}{\partial q_{\mathrm{k}}}\frac{\partial\psi\delta\psi}{\partial q_{\mathrm{k}}} = \int \mathbf{dq}\,\mathrm{div}\,\mathrm{grad}\,\psi\delta\psi = \oint_{\mathrm{S}} \mathbf{dS}\cdot\mathrm{grad}\,\psi\delta\psi = 0,$$

because $\delta\psi = 0$ on the boundary. We are then left with the integral

$$\delta I_{\mathrm{gen}} = -2\int \mathbf{dq}\left[\frac{\hbar^2}{2m}\sum_{\mathrm{j,k}}^{\mathrm{N}} a_{\lambda\mu}\frac{\partial^2\psi}{\partial q_{\mathrm{j}}\partial q_{\mathrm{k}}} + (\mathcal{E} - V(r))\,\psi\right]\delta\psi = 0. \tag{B.21}$$

For arbitrary $\delta\psi$ within the volume, this requires that

$$\frac{\hbar^2}{2m}\sum_{\mathrm{j,k}}^{\mathrm{N}} a_{\lambda\mu}\frac{\partial^2\psi}{\partial q_{\mathrm{j}}\partial q_{\mathrm{k}}} + (\mathcal{E} - V(q))\,\psi = 0, \tag{B.22}$$

within the volume.

Appendix C
Schwarz Inequality

Theorem 1 *For two vectors $|\Psi\rangle$ and $|\Phi\rangle$ in Hilbert space (defined as a complete metric space with a scalar product ([128], p. 296; [81], p. 38)),*

$$|\langle\Psi\,|\Phi\rangle| \leq |\langle\Psi\,|\Psi\rangle|^{1/2}\,|\langle\Phi\,|\Phi\rangle|^{1/2}\,, \tag{C.1}$$

where $\langle\Psi\,|\Phi\rangle$ is the scalar product of $|\Psi\rangle$ and $|\Phi\rangle$.

Proof The theorem is trivially true if either $|\Psi\rangle$ or $|\Phi\rangle$ is the zero vector. If neither $|\Psi\rangle$ or $|\Phi\rangle$ is the zero vector, we have for any scalar α, since $\alpha\,|\Psi\rangle \rightarrow \langle\Psi|\,\alpha^*$,

$$\begin{aligned} 0 \leq Q &= \langle\Psi + \alpha\Phi\,|\Psi + \alpha\Phi\rangle \\ &= \langle\Psi\,|\Psi\rangle + \alpha\,\langle\Psi\,|\Phi\rangle + \alpha^*\,\langle\Phi\,|\Psi\rangle + |\alpha|^2\,\langle\Phi\,|\Phi\rangle\,, \end{aligned} \tag{C.2}$$

where the asterisk denotes the complex conjugate. If this is true for any complex α, then it must be true for the α that minimizes the expression on the right-hand side. Choosing $\alpha = x + iy$ and writing $\langle\Psi\,|\Phi\rangle = a + ib$, Q becomes

$$\begin{aligned} Q &= \langle\Psi\,|\Psi\rangle + (x + iy)\,(a + ib) + (x - iy)\,(a - ib) + \left(x^2 + y^2\right)\langle\Phi\,|\Phi\rangle \\ &= \langle\Psi\,|\Psi\rangle + 2ax - 2by + \left(x^2 + y^2\right)\langle\Phi\,|\Phi\rangle\,. \end{aligned} \tag{C.3}$$

If Q is a minimum, the partial derivatives of Q with respect to x and y must vanish, whence

$$\begin{aligned} \frac{\partial Q}{\partial x} &= 2a + 2x\,\langle\Phi\,|\Phi\rangle = 0\,, \\ \frac{\partial Q}{\partial y} &= -2b + 2y\,\langle\Phi\,|\Phi\rangle = 0\,. \end{aligned} \tag{C.4}$$

C. S. Helrich, *The Quantum Theory—Origins and Ideas*, History of Physics,
https://doi.org/10.1007/978-3-030-79268-8

Then,

$$x = -\frac{a}{\langle \Phi \, | \Phi \rangle} \,,$$
$$y = \frac{b}{\langle \Phi \, | \Phi \rangle} \,. \tag{C.5}$$

The complex quantity α is thus

$$\alpha = -\frac{a - ib}{\langle \Phi \, | \Phi \rangle} = -\frac{\langle \Phi \, | \Psi \rangle}{\langle \Phi \, | \Phi \rangle} . \tag{C.6}$$

Since

$$\langle \Psi \, | \Phi \rangle = a + ib \,,$$
$$\langle \Phi \, | \Psi \rangle = a - ib \,. \tag{C.7}$$

Equation (C.2) becomes

$$0 \leq \langle \Psi \, | \Psi \rangle - \frac{\langle \Phi \, | \Psi \rangle}{\langle \Phi \, | \Phi \rangle} \langle \Psi \, | \Phi \rangle - \frac{\langle \Psi \, | \Phi \rangle}{\langle \Phi \, | \Phi \rangle} \langle \Phi \, | \Psi \rangle + \frac{\langle \Phi \, | \Psi \rangle \langle \Psi \, | \Phi \rangle}{\langle \Phi \, | \Phi \rangle} , \tag{C.8}$$

or

$$\langle \Phi \, | \Psi \rangle \langle \Psi \, | \Phi \rangle \leq \langle \Psi \, | \Psi \rangle \langle \Phi \, | \Phi \rangle \,. \tag{C.9}$$

Taking the positive square root of both sides of (C.9) yields

$$|\langle \Psi \, | \Phi \rangle| \leq |\langle \Psi \, | \Psi \rangle|^{1/2} \, |\langle \Phi \, | \Phi \rangle|^{1/2} \,, \tag{C.10}$$

and the theorem is proven.

Appendix D
Heisenberg's Principle

Here we consider the operator form of the Heisenberg indeterminacy principle. This is mathematically the most general form of the principle. Our mathematical derivation here is for any pair of operators, whether or not they have physical interpretations.

We consider two operators **A** and **B** which satisfy the commutation relationship

$$\mathbf{AB} - \mathbf{BA} = i\hbar\mathbf{1}\,. \tag{D.1}$$

This would be the case for the momentum and position operators, for example.

We define the quantum mechanical average of the operators **A** and $\mathbf{A}^2$ by

$$\begin{aligned} \langle \mathbf{A} \rangle &= \langle \Psi | \, \mathbf{A} \, | \Psi \rangle \\ \left\langle \mathbf{A}^2 \right\rangle &= \langle \Psi | \, \mathbf{A}^2 \, | \Psi \rangle \,, \end{aligned} \tag{D.2}$$

where $|\Psi\rangle$ is the quantum mechanical state vector, which is the general vector form of the wave function. If we are working in a space of eigenvectors $|\mu\rangle$, then $|\Psi\rangle$ takes the form

$$|\Psi\rangle = \sum \Psi_\mu \, |\mu\rangle\,. \tag{D.3}$$

If the eigenvectors $|\mu\rangle$ are represented in spatial coordinates, this becomes

$$\Psi\,(r) = \sum \Psi_\mu \mu\,(r)\,, \tag{D.4}$$

which is the wave function.

We know from the statistical treatment of Max Born that the quantum mechanical average $\langle \mathbf{A} \rangle = \langle \Psi | \, \mathbf{A} \, | \Psi \rangle$ is to be interpreted as the probability density for the value of **A**. We may then define the fluctuation of the operator **A** by

$$\boldsymbol{\delta A} = \mathbf{A} - \langle \mathbf{A} \rangle\,. \tag{D.5}$$

C. S. Helrich, *The Quantum Theory—Origins and Ideas*, History of Physics,
https://doi.org/10.1007/978-3-030-79268-8

The quantum mechanical average of δA vanishes, which is consistent with Born's interpretation. However, we can define an indeterminacy of the physical quantity defined by the operator **A** by

$$\Delta \mathbf{A} = \sqrt{\langle \mathbf{A}^2 \rangle - \langle \mathbf{A} \rangle^2} = \sqrt{\langle \delta A^2 \rangle}. \tag{D.6}$$

The quantity $\Delta \mathbf{A}$ is not zero and has real physical meaning, because both $\langle \mathbf{A}^2 \rangle$ and $\langle \mathbf{A} \rangle^2$ are mathematically defined quantities. We shall call $\Delta \mathbf{A}$ the *indeterminacy* of the physical operator **A**. It is a measure of the spread in the values we would expect to result in our experimental measurement of **A**. In statistical terms, this is the square root of the average of the square of the deviation of the values we may expect between actual measurements of **A** and the probability density predicted for those values in Born's terminology. In a certain sense, this can also be thought of as our uncertainty in the outcome of a measurement.

We note that, for two operators **A** and **B**,

$$\begin{aligned} \delta A \delta B &= (\mathbf{A} - \langle \mathbf{A} \rangle)(\mathbf{B} - \langle \mathbf{B} \rangle) \\ &= \mathbf{AB} - \langle \mathbf{A} \rangle \mathbf{B} - \mathbf{A} \langle \mathbf{B} \rangle + \langle \mathbf{A} \rangle \langle \mathbf{B} \rangle . \end{aligned} \tag{D.7}$$

Since the average values $\langle \mathbf{A} \rangle$ and $\langle \mathbf{B} \rangle$, which are real numbers, commute with all operators and with one another,

$$\delta A \delta B - \delta B \delta A = \mathbf{AB} - \mathbf{BA} = i\hbar \mathbf{1} . \tag{D.8}$$

Then from (D.6) and the Schwarz inequality (§C),

$$(\Delta \mathbf{A})^2 (\Delta \mathbf{B})^2 = \langle \delta A^2 \rangle \langle \mathbf{B}^2 \rangle \geq |\langle \delta A \delta B \rangle|^2 . \tag{D.9}$$

We note that

$$\begin{aligned} \delta A \delta B &= \frac{\delta A \delta B + \delta B \delta A}{2} + \frac{\mathbf{AB} - \delta B \delta A}{2} \\ &= \frac{\delta A \delta B + \delta B \delta A}{2} + \frac{i\hbar}{2} \mathbf{1}. \end{aligned} \tag{D.10}$$

Taking the average of both sides of (D.10), we obtain

$$\langle \delta A \delta B \rangle = \left\langle \frac{\delta A \delta B + \delta B \delta A}{2} \right\rangle + \frac{i\hbar}{2} .$$

Therefore, using (D.9),

$$(\Delta\mathbf{A})^2\,(\Delta\mathbf{B})^2 \geq |\langle\delta\boldsymbol{A}\delta\boldsymbol{B}\rangle|^2 = \left|\left\langle\frac{\delta\boldsymbol{A}\delta\boldsymbol{B}+\delta\boldsymbol{B}\delta\boldsymbol{A}}{2}\right\rangle + \frac{i\hbar}{2}\mathbf{1}\right|^2$$
$$\geq \frac{\hbar^2}{4}\,.$$

Hence, whenever we have the condition

$$\mathbf{AB}-\mathbf{BA}=i\hbar\mathbf{1}\,,$$

it follows mathematically that

$$(\Delta\mathbf{A})\,(\Delta\mathbf{B}) \geq \frac{\hbar}{2}\,. \qquad \text{(D.11)}$$

Appendix E
Displacement Operator

If we consider the quantum mechanical state of a system as a vector $|P\rangle$, the Hamiltonian, which operates on the state vector, is a matrix. Dirac called the vector $|P\rangle$, which has all the properties of any vector, a "ket vector." We then have a way to treat the concepts of the matrix theory developed systematically by Born, Heisenberg, and Jordan. Operators become square matrices and vectors are column matrices.

For every vector $|P\rangle$, there is a dual vector $\langle P|$. The scalar product of two vectors $|A\rangle$ and $|B\rangle$ is $\langle A|\ B\rangle$. If $|P\rangle$ is a column vector, the dual $\langle P|$ is formed by interchanging rows for columns and taking the complex conjugate of the result.

We can displace the state vector $|P\rangle$ through the action of a linear operator $\mathbf{D}$. If the original ket is $|P\rangle$, the displaced ket $|P_{\mathrm{d}}\rangle$ is

$$|P_{\mathrm{d}}\rangle = \mathbf{D}\,|P\rangle\,. \tag{E.1}$$

The dual vector $\langle P_{\mathrm{d}}|$ is then

$$\langle P_{\mathrm{d}}| = \langle P|\,\mathbf{D}^{+}, \tag{E.2}$$

where $\mathbf{D}^{+}$ is the adjoint (transpose of the complex conjugate) of $\mathbf{D}$. For our purposes we do not want the magnitude of the vector $|P\rangle$ to change during the displacement, so we require that

$$\langle P|\ P\rangle = \langle P_{\mathrm{d}}|\ P_{\mathrm{d}}\rangle = \langle P|\,\mathbf{D}^{+}\mathbf{D}\,|P\rangle\ ,$$

or

$$\mathbf{D}^{+}\mathbf{D} = \mathbf{D}\mathbf{D}^{+} = \mathbf{1}\,. \tag{E.3}$$

Therefore, $\mathbf{D}$ is unitary.

C. S. Helrich, *The Quantum Theory—Origins and Ideas*, History of Physics,
https://doi.org/10.1007/978-3-030-79268-8

Matrix operators will also be affected by a translation. Suppose the operator $\mathbf{Q}$ operates on the vector $|P\rangle$ to produce $|R\rangle$. Then,

$$\mathbf{Q}\,|P\rangle = |R\rangle \ .$$

If we denote the displaced operator by $\mathbf{Q}_\mathrm{d}$, the translated operation is

$$\mathbf{Q}_\mathrm{d}\,|P_\mathrm{d}\rangle = |R_\mathrm{d}\rangle \ . \tag{E.4}$$

Using (E.1), Eq. (E.4) becomes

$$\mathbf{Q}_\mathrm{d}\,|P_\mathrm{d}\rangle = \mathbf{D}\,|R\rangle = \mathbf{DQ}\,|P\rangle = \mathbf{DQD}^{+}\,|P_\mathrm{d}\rangle \ .$$

Hence,

$$\mathbf{Q}_\mathrm{d} = \mathbf{DQD}^{+} \tag{E.5}$$

for the general displacement of the operator $\mathbf{Q}$. We note that (E.5) is still valid if $\mathbf{D}$ includes an arbitrary complex exponential factor $\exp(i\gamma)$, where γ is real. Hence, $\mathbf{D}$ is defined to within a phase factor $\exp(i\gamma)$.

From the continuity of the displacement of physical vectors such as $|P\rangle$, we require existence of the following limit:

$$\lim_{\delta x \to 0} \frac{|P_\mathrm{d}\rangle - |P\rangle}{\delta x} = \lim_{\delta x \to 0} \frac{\mathbf{D} - \mathbf{1}}{\delta x}\,|P\rangle = \mathbf{d}_\mathrm{x}\,|P\rangle \ . \tag{E.6}$$

We call this limit the *displacement operator* $\mathbf{d}_x$. Hence,

$$\mathbf{d}_x = \lim_{\delta x \to 0} \frac{\mathbf{D} - \mathbf{1}}{\delta x}. \tag{E.7}$$

To within an additive imaginary constant and to first order in δx, $\mathbf{D}$ is then

$$\mathbf{D} = \mathbf{1} + \delta x \mathbf{d}_x. \tag{E.8}$$

If we now use (E.3), we have, to first order in δx,

$$\begin{aligned} \mathbf{1} = \mathbf{D}^{+}\mathbf{D} &= \left(\mathbf{1} + \delta x \mathbf{d}_x^{+}\right)\left(\mathbf{1} + \delta x \mathbf{d}_x\right) \\ &= \mathbf{1} + \delta x \left(\mathbf{d}_x^{+} + \mathbf{d}_x\right). \end{aligned} \tag{E.9}$$

That is,

$$\mathbf{d}_x^{+} = -\mathbf{d}_x \ , \tag{E.10}$$

or $(d_x)_\mathrm{jk} = -\,(d_x)^*_\mathrm{kj}$ for all elements. This can only be true if the elements are pure imaginary. Therefore, $\mathbf{d}_x$ is itself (mathematically) a purely imaginary operator.

To first order in δx, a displaced dynamical variable $\mathbf{Q}$ becomes, using (E.5),

$$\begin{aligned}\mathbf{Q}_{\mathrm{d}} &= (\mathbf{1} + \delta x \mathbf{d}_x)\,\mathbf{Q}\left(\mathbf{1} + \delta x \mathbf{d}_x^{+}\right)\\ &= (\mathbf{1} + \delta x \mathbf{d}_x)\,\mathbf{Q}\,(\mathbf{1} - \delta x \mathbf{d}_x)\\ &= \mathbf{Q} + \delta x\,(\mathbf{d}_x\mathbf{Q} - \mathbf{Q}\mathbf{d}_x)\,. \end{aligned} \tag{E.11}$$

Hence,

$$\lim_{\delta x \to 0} \frac{\mathbf{Q}_{\mathrm{d}} - \mathbf{Q}}{\delta x} = (\mathbf{d}_x\mathbf{Q} - \mathbf{Q}\mathbf{d}_x)\ , \tag{E.12}$$

which is the definition of a *derivative* of the operator $\mathbf{Q}$.

Now we suppose that we have a piece of apparatus that measures the distance x on the x-axis. We call the operator representing the action of this apparatus $\mathbf{x}$. Using the bold lowercase notation here for an operator is not consistent with our previous use of the uppercase letters for operators. We do this, however, for comparison with the notation of the Göttingen matrix mechanics.

We translate this apparatus a short distance δx down the x-axis, using the operator $\mathbf{D}$ in the form (E.8). Using the transformation of operators (E.11), the operator $\mathbf{x}_{\mathrm{d}}$ is

$$\mathbf{x}_{\mathrm{d}} = \mathbf{x} + \delta x\,(\mathbf{d}_x\mathbf{x} - \mathbf{x}\mathbf{d}_x)\,. \tag{E.13}$$

From (E.12), the derivative of the operator $\mathbf{x}$ is

$$\lim_{\delta x \to 0} \frac{\mathbf{x}_{\mathrm{d}} - \mathbf{x}}{\delta x} = (\mathbf{d}_x\mathbf{x} - \mathbf{x}\mathbf{d}_x)\,. \tag{E.14}$$

Applying (E.14) to the vector $|X\rangle$ results in

$$\begin{aligned}\lim_{\delta x \to 0} \frac{\mathbf{x}_{\mathrm{d}}\,|X\rangle - \mathbf{x}\,|X\rangle}{\delta x} &= \lim_{\delta x \to 0} \frac{(x - x)\,|X\rangle - x\,|X\rangle}{\delta x}\\ &= \lim_{\delta x \to 0} \left(-\frac{\delta x}{\delta x}\right)|X\rangle\\ &= (\mathbf{d}_x\mathbf{x} - \mathbf{x}\mathbf{d}_x)\,|X\rangle\ , \end{aligned} \tag{E.15}$$

since the displaced operator $\mathbf{x}_{\mathrm{d}}$ returns the value $x - \delta x$. Therefore,

$$\mathbf{d}_x\mathbf{x} - \mathbf{x}\mathbf{d}_x = -\mathbf{1}\,. \tag{E.16}$$

Multiplying (E.16) by $i\hbar$ (with $\hbar = h/2\pi$), we have

$$i\hbar\mathbf{d}_x\mathbf{x} - i\hbar\mathbf{x}\mathbf{d}_x = -i\hbar\mathbf{1}. \tag{E.17}$$

The Eq. (E.17) is of the same form as Eq. (7.8), which we repeat here for convenience:

$$\mathbf{p}_x\mathbf{x} - \mathbf{x}\mathbf{p}_x = -i\hbar\mathbf{1}\,. \tag{E.18}$$

In (E.18), we have the coordinate as $\mathbf{x}$ and the momentum as $\mathbf{p}_x$, to correspond to the situation we are considering.

Subtracting (E.17) from (E.18), we have

$$(\mathbf{p}_x - i\hbar\mathbf{d}_x)\,\mathbf{x} - \mathbf{x}\,(\mathbf{p}_x - i\hbar\mathbf{d}_x) = \mathbf{0}. \tag{E.19}$$

That is, $(\mathbf{p}_x - i\hbar\mathbf{d}_x)$ commutes with $\mathbf{x}$. If we now consider $\mathbf{Q}$ to be any dynamical operator unaffected by the translation δx, we find that

$$(\mathbf{p}_x - i\hbar\mathbf{d}_x)\,\mathbf{Q} - \mathbf{Q}\,(\mathbf{p}_x - i\hbar\mathbf{d}_x) = \mathbf{0}. \tag{E.20}$$

Therefore, the operator $(\mathbf{p}_x - i\hbar\mathbf{d}_\mathrm{x})$ commutes with $\mathbf{x}$ and with all operators unaffected by translations in x. It must then simply be a number, which we may choose to be zero ([86], p. 103). We conclude that the operator $\mathbf{d}_x$ is [[86], p. 103]

$$i\hbar\mathbf{d}_x = \mathbf{p}_x. \tag{E.21}$$

Following a similar argument for a time displacement results in

$$i\hbar\mathbf{d}_\mathrm{t} = \mathbf{H}\,, \tag{E.22}$$

in place of (E.21).

References

1. https://history.aip.org/history/exhibits/heisenberg/p09.htm. Accessed 24 July 2019
2. APS NEWS **21** (3) (2012)
3. APS NEWS **16**(1) (2007)
4. A. Alexander, Infinitesimal. Scientific American (2014)
5. H.S. Allen, Charles Glover Barkla. 1877–1944. Obituary Notices Fellows R. Soc. **5**(15), 341 (1947) JSTOR. www.jstor.org/stable/769087
6. Aristotle, Physics, Book IV, section 8
7. N.W. Ashcroft, N.D. Mermin, *Solid State Physics* (Holt, Rinehart, and Winston, Philadelphia, 1976)
8. https://en.wikipedia.org/wiki/Johann_Jakob_Balmer. Accessed 6 May 2019
9. C.G. Barkla, C.A. Sadler, Phil. Mag. **17**, 739 (1909)
10. C.G. Barkla, Phil. Mag. **21**, 648 (1911)
11. S. Berryman, "Leucippus", *The Stanford Encyclopedia of Philosophy* (Winter 2016 Edition), ed. by E.N. Zalta. https://plato.stanford.edu/archives/win2016/entries/leucippus/
12. S. Berryman, "Democritus", *The Stanford Encyclopedia of Philosophy* (Winter 2016 Edition), ed. by E.N. Zalta. https://plato.stanford.edu/archives/win2016/entries/democritus/
13. M. Bôcher, *Introduction to Higher Algebra* (Macmillan, New York, 1907)
14. M. Bôcher, *Einführung in Die Höhere Algebra*, Translated by H. Beck (Teubner, Leipzig, 1910)
15. N. Bohr, Phil. Mag. **26**, 1 (1913)
16. N. Bohr, *Collected Works Volume 5, The Emergence of Quantum Mechanics (mainly 1924–1926)*, ed. by K. Stolzenburg (North Holland, Elsevier Science, Amsterdam, 1984)
17. N. Bohr, H.A. Kramers, J.C. Slater, London, Edinburgh, and Dublin Phil. Mag. J. Sci. **47**, 785 (1924)
18. L. Boltzmann, Wien. Ber. **63**, 679 (1871)
19. L. Boltzmann, Wien. Ber. **66**, 275 (1872)
20. L. Boltzmann, Wien. Ber. **76**, 373 (1877). English translation by K. Sharp, F. Matschinsky, Entropy **17**(4), 1971–2009 (2015). https://doi.org/10.3390/e17041971
21. L. Boltzmann, Annalen der Physik un Chemie **258**, 291 (1884)
22. L. Boltzmann, *Vorlesungen über Gastheorie, Parts I and II* (Barth, Leipzig, 1896, 1898) translated as *Lectures on Gas Theory* (University of California Press, Berkeley, 1964, reprint Dover, New York, 1995)

C. S. Helrich, *The Quantum Theory—Origins and Ideas*, History of Physics, https://doi.org/10.1007/978-3-030-79268-8

23. M. Born, P. Jordan, Z. Physik **34**, 858 (1925)
24. M. Born, W. Heisenberg, P. Jordan, Z. Physik **35**, 557 (1926)
25. M. Born, N. Wiener, Z. Physik **36**, 174 (1926)
26. M. Born, Z. Physik **37**, 863 (1926). https://doi.org/10.1007/BF01397477
27. M. Born, Biographical Memoirs of Fellows R. Soc. **6**(17) (01 Nov 1948)
28. M. Born, *Nobel Lecture*. NobelPrize.org. Nobel Media AB 2020. Sunday 26 July 2020. https://www.nobelprize.org/prizes/physics/1954/born/lecture/
29. W. Bothe, H. Geiger, Die Naturwissenschaften **13**, 440 (1925), Z. Physik **32**, 639 (1925)
30. W. Bothe, H. Geiger, *Mitteilung aus der Physikalisch-Technischen Reichsanstalt* (25 April 1925)
31. *Encyclopaedia Britannica*. https://www.britannica.com/topic/Bell-Laboratories
32. W.L. Bragg, Proc. Camb. Phil. Soc. **17**, 43 (1912)
33. *Encyclopaedia Britannica*, vol. 3 (1969)
34. *Encyclopaedia Britannica*, vol. 4 (1969)
35. *Encyclopaedia Britannica*, vol. 11 (1969)
36. *Encyclopaedia Britannica*, vol. 15 (1969)
37. *Encyclopaedia Britannica*, vol. 21 (1969)
38. www.britannica.com/biography/Michael-Faraday. Accessed 19 November 2018
39. B.R. Brown, *Planck: Driven by Vision, Broken by War* (Oxford University Press, Oxford, 2015)
40. S.G. Brush, *Kinetic Theory VI: The Nature of Gases and Heat* (Pergamon Press, Oxford, 1965)
41. S.G. Brush, *Storia della Scienza*, vol. 7, ed. by S. Petruccioli, L'Ottocento, Chapter 44 (2004)
42. S. Bugajski, Int. J. Theoret. Phys. **30**, 961 (1991)
43. C.H.D. Buijs-Ballot, Annalen der Physik **103**, 240 (1859)
44. H. Capellmann, arXiv:1606.00190v2 [Physics.hist-ph] (26 Nov 2016)
45. S. Carnot, *Reflexions sur la puissance motrice du feu et sur les machines propre à développer cette puissance* (Bachelier, Paris, 1824). English translation S. Carnot, *Reflections on the Motive Power of Fire and Other Papers on the Second Law of Thermodynamics*, ed. by E. Mendoza (Dover, New York, 1960)
46. C. Cercignani, *Ludwig Boltzmann: The Man Who Trusted Atoms* (Oxford University Press, Oxford, 1998)
47. A. Chalmers, *Atomism from the 17th to the 20th Century*, The Stanford Encyclopedia of Philosophy (Spring 2019 Edition), ed. by E.N. Zalta. https://plato.stanford.edu/archives/spr2019/entries/atomism-modern/
48. S. Chapman, T.G. Cowling, *The Mathematical Theory of Non-Uniform Gases*, 3rd edn. (Cambridge University Press, London, 1970)
49. R. Clausius, Annalen der Physik **79**(4), 368, 500 (1850)
50. R. Clausius, Annalen der Physik **100**, 353 (1857)
51. R. Clausius, Phil. Mag. **14**, 108 (1857)
52. R. Clausius, Annalen der Physik **105**, 239 (1858)
53. R. Clausius, Phil. Mag. **17**, 81 (1859)
54. R. Clausius, Annalen der Physik **125**, 353 (1865)
55. R. Clausius, *The Mechanical Theory of Heat—with Its Applications to the Steam Engine and to Physical Properties of Bodies* (van Voorst, London, 1865)
56. A.H. Compton, Phys. Rev. **21**(5), 483 (1923)
57. A.H. Compton, *Joint Session of the AIP and AAPT*, New York (Feb 3, 1961)
58. A.H. Compton, S.K. Allison, *X-Rays in Theory and Experiment* (D. Van Nostrand Company, Inc., Princeton, NJ, 1935)
59. A.H. Compton, A.W. Simon, Phys. Rev. **26**, 289 (1925)
60. R. Courant, D. Hilbert, *Methoden Der Mathematischen Physik* (Springer, Berlin, 1924)
61. R. Courant, D. Hilbert, *Methods of Mathematical Physics* (Interscience, New York, 1953)
62. W. Crookes, London, Edinburgh, and Dublin Phil. Mag. J. Sci. **7**, 57 (1879)
63. W.H. Cropper, *Great Physicists* (Oxford University Press, New York, 2001)

64. O. Darrigol, Hist. Stud. Phys. Biol. Sci. **19**(1), 17–80 (1988)
65. E.E. Daub, Isis **62**(4), 512 (1971)
66. C. Davisson, C. Kunsman, Science **54**, 522 (1921)
67. C. Davisson, L. Germer, Nature **119**, 558 (1927)
68. C. Davisson, L. Germer, Bell Lab. Rec. **4**, 257 (1927)
69. C. Davisson, L. Germer, (abstract) Phys. Rev. **29**, 908 (1927)
70. C. Davisson, L. Germer, Phys. Rev. **30**, 705 (1927)
71. L. de Broglie, J. Phys. et rad. **3**, 422 (1922)
72. L. de Broglie, Comptes rendus (Paris) **177**, 507, 548, 630 (1923)
73. L. de Broglie, *Recherches Sur La Théorie Des Quanta* (Thesis, Paris, 1924)
74. L. de Broglie, *Recherches sur la Théorie des Quanta* Ann. de Phys., 10e série, t. III (janvier–février 1925, trans. by A.F. Kracklauer AFK 2004)
75. P. Debye, Physikalische Zeitschrift **14**, 259 (1913)
76. P. Debye, Physikalische Zeitschrift **17**, 507 (1916)
77. P. Debye, Nachr. Ges. Wiss. Gottingen Math. Phys. Klasse **142** (1916)
78. P. Debye, Physikalische Zeitschrift **17**, 512 (1916)
79. P. Debye, Nachr. Ges. Wiss. Gottingen Math. Phys. Klasse **161** (1916)
80. P. Debye, Biog. Memoirs Fellows Roy. Soc. **16**, 175 (Nov., 1970)
81. J.W. Dettman, *Mathematical Methods in Physics and Engineering* (McGraw-Hill, New York, 1962)
82. R.H. Dicke, J.P. Wittke, *Introduction to Quantum Mechanics* (Addison-Wesley, Reading, MA, 1960)
83. P.A.M. Dirac, Proc. Roy. Soc. **109**, 642 (1925)
84. P.A.M. Dirac, Proc. Roy. Soc. **110**, 561 (1925)
85. P.A.M. Dirac, Proc. Roy. Soc. **111**, 281 (1926)
86. P.A.M. Dirac, *The Principles of Quantum Mechanics* (Oxford University Press, Oxford, 1958)
87. P.A.M. Dirac, Nature **189**, 355 (1961)
88. P.A.M. Dirac, *Proceedings of the International School of Physics "Enrico Fermi" Course LVII*, ed. by C. Weiner (Academic Press, New York, 1977)
89. P. Drude, *Annalen der Physik* **1**, 566, and **3**, 369 (1900)
90. R. Dugas, *History of Mechanics* (Dover, Mineola, New York, 1988)
91. M. Eckert, Annalen der Physik **524**(5), A83–A85 (2012)
92. M. Eckert, Sommerfeld School, in *Compendium of Quantum Physics*, ed. by D. Greenberger, K. Hentschel, F. Weinert (Springer, Berlin, Heidelberg, 2009)
93. M. Eckert, Phys. Perspect. **1**, 238 (1999)
94. A. Einstein, Annalen der Physik **14**, 354 (1904)
95. A. Einstein, Annalen der Physik **17**(10), 891 (1905)
96. A. Einstein, Annalen der Physik **17**(6), 132 (1905)
97. A. Einstein, Annalen der Physik **17**(8), 549 (1905)
98. A. Einstein et al., *The Principle of Relativity* (Dover, New York, 1952)
99. A. Einstein, Annalen der Physik **49**, 769 (1916)
100. A. Einstein, Ver. Dtsch. Phys. Ges. **18**, 318 (1916)
101. A. Einstein, Physikalische Zeitschrift **18**, 121 (1917)
102. A. Einstein, B. Podolsky, N. Rosen, Phys. Rev. **47**, 777 (1935)
103. A. Einstein, A. Sommerfeld, *Briefwechsel. Sechzig Briefe aus den goldenen Zeitalter der modernen Physik*, edited and annotated by A. Hermann (Basel-Stuttgart, 1968)
104. P. Ewald, Annalen der Physik **44**, 257 (1914)
105. M. Faraday, *Experimental Researches in Electricity* (Dover, New York, 1965)
106. W.A. Fedak, J.J. Prentis, Am. J. Phys. **70**, 332 (2002)
107. W.A. Fedak, J.J. Prentis, Am. J. Phys. **77**, 128 (2009)
108. G.F. FitzGerald, R.J.E. Clausius, Obituary Notices Proc. Roy. Soc. London **48** (1890)
109. P. Forman, Arch. Hist. Exact Sci. **6**, 38 (1969)
110. P. Forman, A. Hermann, *Sommerfeld, Arnold (Johannes Wilhelm)*, Encyclopedia.com Complete Dictionary of Scientific Biography (Charles Scribner's Sons, 2008)

111. R. Fox, *The Caloric Theory of Gases: From Lavoisier to Regnault* (Clarendon Press, Oxford, 1971)
112. R. Fox, *Sadi Carnot: Reflexions on the Motive Power of Fire, a Critical Edition with the Surviving Manuscripts* (Manchester University Press, Manchester, 1986)
113. P. Frame, orau.org/ptp/collection/electrometers/quadrantelectrometer.htm. Accessed 12 2020
114. J. Franck, G. Hertz, Ver. d. Deutsch. Phys. Ges. **16**, 457 (1914)
115. J. Franck, G. Hertz, Ver. d. Deutsch. Phys. Ges. **16**, 512 (1914)
116. J. Franck, G. Hertz, Phys. Zs. **17**, 132 (1919)
117. B. Friedrich, D. Herschbach, Daedalus **127**(1), 165 (1998)
118. B. Friedrich, D. Herschbach, Phys. Today **56**(12), 53 (2003). https://doi.org/10.1063/1.1650229
119. W. Friedrich, P. Knipping, Sitz.ber. Bayer. Akad. Wiss. 311–322 (1912)
120. G. Galilei, Dialogo Dei Due Massimi Sistemi Del Mondo [Dialogues Concerning the Two Chief World Systems], Translated by D. Stillman (University of California Press, Los Angeles, 1967)
121. R.K. Gehrenbeck, Phys. Today (January, 1978)
122. H. Geiger, E. Marsden, Proc. R. Soc. **82**, 495 (1909)
123. J.W. Gibbs, *Elementary Principles in Statistical Mechanics* (Yale University Press, Hartford 1902, reprint Dover, New York, 1960)
124. E. Goldstein, Monatsberichte der Königlich Preussischen Akademie der Wissenschaften zu Berlin **279** (1876)
125. A. De Gregorio, Stud. Hist. Phil. Modern Phys. **45**, 72 (2014)
126. W. Greiner, *Quantum Mechanics, An Introduction* (Springer, Berlin, Heidelberg, 1989)
127. J. Gribbin, *Erwin Schrödinger and the Quantum Revolution* (Wiley, Hoboken, NJ, 2013)
128. S. Hassani, *Foundations of Mathematical Physics* (Allyn and Bacon, 1991)
129. J.L. Heilbron, *Proceedings of the International School of Physics "Enrico Fermi" Course LVII*, ed. by C. Weiner (Academic Press, New York, 1977)
130. J.L. Heilbron, *The Dilemmas of an Upright Man* (Harvard University Press, 2000)
131. W. Heisenberg, Biographical. NobelPrize.org
132. W. Heisenberg, *The Physical Principles of the Quantum Theory* (University of Chicago Press, Chicago, 1930)
133. W. Heisenberg, Z. Physik **43**, 172 (1927)
134. W. Heisenberg, P. Jordan, Z. Physik **37**, 263 (1926)
135. W. Heisenberg, Z. Physik **33**, 879 (1925)
136. W. Heisenberg, *Der Teil und das Ganze. Gespräche im Umkreis der Atomphysik* (R. Piper & Co. Verlag, Munich, 1969)
137. H. Helmholtz, *Die Erhaltung Der Kraft, Eine Physikalische Abhandlung* (Druck und Verlag G, Reimer Berlin, 1847)
138. C.S. Helrich, *Modern Thermodynamics with Statistical Mechanics* (Springer, Berlin, Heidelberg, 2009)
139. C.S. Helrich, *The Classical Theory of Fields: Electromagnetism* (Springer, Berlin, Heidelberg, 2012)
140. C.S. Helrich, *Analytical Mechanics* (Springer, Berlin, Heidelberg, 2016)
141. G. Hettner, Die Naturwissenschaften **48**, 1033 (1922)
142. R.G. Hewlett, O.E. Anderson, *The New World* (University of California Press, Berkeley, 1990)
143. G. Holton, S.G. Brush, *Physics the Human Adventure*, 3rd edn. (Rutgers University Press, New Brunswick, New Jersey, 2001)
144. G. Holton, Phys. Today (July, 2000)
145. F.A. Homann, Synth. Phil. **4**(2), 557 (1989)
146. C.J. Isham, *Lectures on Quantum Theory* (Imperial College Press, London, 1995)
147. M. Jammer, *Concepts of Mass in Classical and Modern Physics* (Harvard University Press 1961, Reprinted by Dover (Mineola, New York, 1997)
148. M. Jammer, *The Conceptual Development of Quantum Mechanics* (American Institute of Physics, Tomas, 1989)

149. E.T. Jaynes, Am. J. Phys. **33**(5), 391 (1965)
150. H. Ibach, H. Lüth, *Solid-State Physics, An Introduction to Theory and Experiment* (Springer, Berlin, Heidelberg, 1991)
151. J. Jeans, *An Introduction to the Kinetic Theory of Gases* (University Press, Cambridge, 1962)
152. I. Kaplan, *Nuclear Physics* (Addison-Wesley, Reading, 1955)
153. M.F. Kimmitt, J. Biol. Phys. **29**, 77 (2003)
154. G. Kirchhoff, Ann. Phys. Chem. **109**, 275 (1860)
155. G. Kirchhoff, Phil. Mag. **20**(4), 1 (1860)
156. M.J. Klein, Phys. Today (November, 1966)
157. M.J. Klein, *Proceedings of the International School of Physics "Enrico Fermi" Course LVII*, ed. by C. Weiner (Academic Press, New York, 1977)
158. D. Kondepudi, I. Prigogine, *Modern Thermodynamics* (Wiley, New York, 1998)
159. H. Kragh, Phys. Teach. **35**(6), 328 (1997)
160. H. Kragh. Phys. World (2000)
161. H. Kragh, J.M. Overduin, Planck's second quantum theory, in *The Weight of the Vacuum* (Springer Briefs in Physics. Springer, Berlin, Heidelberg, 2014)
162. H.A. Kramers, Nature (May 10, 1924)
163. H.A. Kramers, W. Heisenberg, Z. Physik **31**, 681 (1925)
164. A.K. Krönig, Poggendorff's Annalen, **99**, 315 (1856), later Annalen der Physik, **33**, 315 (1856)
165. C. Lanczos, *The Variational Principles of Mechanics* (Dover, Mineola, New York, 1986)
166. Laplace, *A Philosophical Essay on Probabilities*, Trans. F.W. Truscott, F.L. Emory (John Wiley and Sons, London, 1902)
167. Laplace, *Traité de Mecanique Céleste* (Duprat, Paris, 1798–1825)
168. M. von Laue, Annalen der Physik **18**(4), 523 (1905)
169. A. Lavoisier, *Elements of Chemistry*, trans. from *Traité Élémentaire de Chimie* by R. Kerr (Dover, Mineola, New York, 1965)
170. P. Lenard, Annalen der Physik **287**(2), 225, **288d**(5), 23 (1894)
171. P. Lenard, Annalen der Physik **8**, 149 (1902)
172. E.v. Lommel, Wied. Ann. **3**, 251 (1877)
173. Lord Rayleigh, London, Edinburgh, and Dublin Phil. Mag. J. Sci. **49**, 539 (1900)
174. O. Lummer, F. Kurlbaum, Verh. Deutsch. Phys. Ges. **17**, 106 (1898)
175. O. Lummer, E. Pringsheim, Verh. Deutsch. Phys. Ges. **2**, 163 (1900)
176. C.F. Manara, Istit. Lombardo Accad. Sci. Lett. Rend. A **123**, 215 (1989)
177. H. Margenau, G.M. Murphy, *The Mathematics of Physics and Chemistry*, vol. 2 (Van Nostrand, New York, 1964)
178. I. Martinson, L.J. Curtis, Nuclear instruments & methods in physics research. Section B: beam interactions with materials and atoms **235**(1–4), 17 (2005)
179. J.C. Maxwell, Phil. Mag. **19**, 19 (1860)
180. J.C. Maxwell, Phil. Mag. **20**, 21 (1860)
181. J.C. Maxwell, Phil. Trans. R. Soc. London **155**, 459 (1865)
182. J.C. Maxwell, On the dynamical theory of gases, in *The Scientific Papers of James Clerk Maxwell*, vol. 2, ed. by D. Niven (Dover, New York, 1965)
183. A.M. Mayer, Nature **18**, 258 (1878)
184. J. Mehra, H. Rechenberg, *The Historical Development of Quanrum Theory*, vol. 1, Part 1 (Springer, New York, 1982)
185. J. Mehra, H. Rechenberg, *The Historical Development of Quantum Theory*, vol. 1, Part 2 (Springer, New York, 1982)
186. J. Mehra, H. Rechenberg, *The Historical Development of Quantum Theory*, vol. 2 (Springer, New York, 1982)
187. J. Mehra, H. Rechenberg, *The Historical Development of Quantum Theory*, vol. 3 (Springer, New York, 1982)
188. J. Mehra, *The Golden Age of Theoretical Physics*, vol. 1 (World Scientific, Singapore, 2001)
189. E. Merzbacher, *Quantum Mechanics*, 3rd edn. (Wiley, New York, 1998)

190. A. Messiah, *Quantum Mechanics* (Dover, New York, 1999)
191. W. Michelson, Journal de Phys. Théor. et Appl. **6**(1), 467 (1887)
192. R.A. Millikan, Phys. Rev. **2**, 109 (1913)
193. R.A. Millikan, Phys. Rev. **7**, 355 (1916)
194. R.A. Millikan, *The Electron: Its Isolation and Measurement and the Determination of Some of its Properties* (University of Chicago Press, Chicago, 1917)
195. R.A. Millikan, Biographical. NobelPrize.org. Nobel Media AB 2020. Fri. 10 Jul 2020. https://www.nobelprize.org/prizes/physics/1923/millikan/biographical/
196. R.A. Millikan, *The Autobiography of Robert A. Millikan* (Prentice-Hall, New York, 1950)
197. A. Einstein et al., *The Principle of Relativity* (Dover, New York, 1952)
198. H. Minkowski, Nachr. Ges. Wiss. Götingen, pp. 53–111 (21 December 1907), reprinted in Mathematiche Annalen, **68**, 472 (1910)
199. D.C. Montgomery, D.A. Tidman, *Plasma Kinetic Theory* (McGraw-Hill, New York, 1964)
200. W.J. Moorer, *Schrödinger: Life and Thought* (Cambridge University Press, Cambridge, 1989)
201. H.G.J. Moseley, Phil. Mag. **26**, 1024 (1913)
202. H.G.J. Moseley, Phil. Mag. **27**, 703 (1914)
203. F. Müller, Br. J. Hist. Sci. **44**(2), 211 (2011)
204. H. Nagaoka, Phil. Mag. **7**, 445 (1904)
205. Nature **141**, 352 (1938)
206. J. von Neumann, *Mathematical Foundations of Quantum Mechanics*, Translated from the German by R.T. Beyer (Princeton University Press, Princeton, 1955)
207. E.F. Nichols, *Berliner Berichte* (November 5, 1896)
208. E.F. Nichols, Phys. Rev. **4**, 297 (1897)
209. J.W. Nicholson, Month. Not. Roy. Ast. Soc. **72**, 49, 176, 693 (1911/12)
210. J.W. Nicholson, Nature **92**, 199 (1913)
211. "The Nobel Prize in Physics 1954". Nobelprize.org. Nobel Media AB 2014. Web. 2 Mar 2017. http://www.nobelprize.org/nobel_prizes/physics/laureates/1954/
212. Ernest Rutherford—Facts. NobelPrize.org. Nobel Media AB 2019. Wed. 1 2019. https://www.nobelprize.org/prizes/chemistry/1908/rutherford/facts/
213. B. Ørsted, Review of Huygens' principle and hyperbolic equations, by Paul Günther. Bull. Am. Math. Soc. **3**(1) (1990)
214. *Oxford Dictionary of Scientists* (Oxford University Press, Oxford, 1999)
215. A. Pais, *Subtle Is the Lord* (Oxford University Press, Oxford, 1982)
216. J. Palmer, *"Parmenides", The Stanford Encyclopedia of Philosophy (Winter 2020 Edition)*, ed. by E.N. Zalta, forthcoming. https://plato.stanford.edu/archives/win2020/entries/parmenides
217. D. Park, *The How and the Why* (Princeton University Press, Princeton, 1988)
218. F. Paschen, Annalen der Physik **60**, 662 (1897)
219. F. Paschen, Annalen der Physik **50**, 901 (1916)
220. W. Pauli, Z. Physik **36**, 336 (1926)
221. W. Pauli, Science **103**(2669), 213 (1946)
222. M. Planck, Sitzungsbericht Preuss. Akad. Wiss. **440** (1899)
223. M. Planck, Annalen der Physik **1**, 99 (1900)
224. M. Planck, Ver. Dtsch. Phys. Ges. **2**(13), 202 (1900)
225. M. Planck, Ver. Dtsch. Phys. Ges. **2**(17), 237 (1900)
226. M. Planck, Annalen der Physik **4**(4), 553 (1901)
227. M. Planck, *Vorlesungen Über Die Theorie Der Wärmestrahlungen* (Barth, Leipzig, 1906, 1913)
228. M. Planck, Nobel Prize Address, in *A Survey of Physical Theory*, p. 102 (Dover, New York 1960)
229. Max Planck—Biographical. NobelPrize.org. Nobel Media AB 2020. Mon. 1 Jun 2020. https://www.nobelprize.org/prizes/physics/1918/planck/biographical/
230. M. Planck to R.W. Wood, 7 October 1931. This letter is part of the collection in the Archives of the Center for the History and Philosophy of Physics of the American Institute of Physics in New York City

231. W. Prout, Ann. Phil. **6**, 321 (1815)
232. PTB Info Sheet—PTR and PTB: History of an Institution, https://www.ptb.de/cms/fileadmin/internet/presse_aktuelles/broschueren/geschichte_ptb/PTR_and_PTB_History_of_an_Institution.pdf. Accessed 22 2020
233. C.W. Ramsauer, Annalen der Physik, **45** , 1120, and **45**, 961 (1914)
234. J.W. Strutt (Lord Rayleigh), Phil. Mag. **49**, 539 (1900)
235. L.E. Reichl, *A Modern Course in Statistical Physics*, 2nd edn. (Wiley, New York, 1998)
236. S. Reif-Acherman, Proc. IEEE **103**, 1672 (2015)
237. W. Ritz, Physikalische Zeitschrift **9**, 521 (1908)
238. R. Rhodes, *The Making of the Atomic Bomb* (Simon and Schuster, New York, 1986)
239. H. Rubens, F. Kurlbaum, Sitzungsbericht der Akad. Wiss. Berlin, Oct 25, 1900
240. A. Russo, Hist. Stud. Phys. Sci. **12**(1), 117 (1981)
241. E. Rutherford, Phil. Mag. **21**(6), 669 (1911)
242. J.R. Rydberg, *Kongliga Svenska Vetenskaps-Akademiens Handlingar* (in French), **23** (11), 1 (1889), English summary: J.R. Rydberg, London, Edinburgh, and Dublin Phil. Mag. J. Sci. **29**, 331 (1890)
243. E. Segrè, Biogr. Mem. Natl. Acad. Sci. **43**, 215 (1973)
244. S. Seth, *Crafting the Quantum: Arnold Sommerfeld and the Practice of Theory, 1890–1926* (MIT Press, Cambridge, 2010)
245. B.L. Silver, *The Ascent of Science* (Oxford University Press, Oxford, 1998)
246. A. Sommerfeld, Sitzb. Münch. Ak. **425** (1915)
247. A. Sommerfeld, Annalen der Phys **51**, 1 (1916)
248. A. Sommerfeld, *Thermodynamics and Statistical Mechanics* (Academic Press, NY, 1956)
249. E. Schrödinger, Annalen der Phys **79**, 361 (1926)
250. E. Schrödinger, Annalen der Phys **79**, 489 (1926)
251. E. Schrödinger, Annalen der Phys **79**, 734 (1926)
252. E. Schrödinger, Annalen der Phys **80**, 437 (1926)
253. E. Schrödinger, Annalen der Phys **81**, 109 (1926)
254. E. Schrödinger, Phys. Rev. **28**, 1049 (1926)
255. A. Schuster, Proc. R. Soc. **37**, 317 (1884)
256. W. Scott, *The Conflict Between Atomism and Conservation Theory 1644–1860* (MacDonald, London, 1970)
257. I. Stakgold, *Green's Functions and Boundary Value Problems* (Wiley, New York, 1979)
258. J. Stefan, Sitzungsberichte der Mathematisch-naturwissenschaftlichen Classe der Kaiserlichen Akademie der Wissenschaften, **79**, 391 (1879)
259. O. Stern—Biographical. NobelPrize.org. *Nobel Lectures, Physics 1942–1962* (Elsevier Publishing Company, Amsterdam, 1964)
260. G.J. Stoney, London, Edinburgh, and Dublin Phil. Mag. J. Sci. **38**, 418 (1894)
261. R.H. Stuewer, Paper presented at the HQ-1 Conference on the History of Quantum Physics at the Max Planck Institute for the History of Science, Berlin, Germany, July 5, 2007
262. J. Suzuki, *Mathematics in Historical Context* (Mathematical Association of America, 2009)
263. M.J. Taltavull, Annalen der Phys (Berlin) **530** (2018)
264. D. ter Haar, *The Old Quantum Theory* (Pergamon Press, Oxford, 1967)
265. M. Thiesen, Ver. Dtsch. Phys. Ges. **2**, 65 (1900)
266. J.J. Thomson, London, Edinburgh, and Dublin Phil. Mag. J. Sci. **38**, 358 (1894)
267. J.J. Thomson, Electrician **39**, 104 (1897)
268. J.J. Thomson, London, Edinburgh, and Dublin Phil. Mag. J. Sci. **44**, 298 (1897)
269. J.J. Thomson, London, Edinburgh, and Dublin Phil. Mag. J. Sci. **7**, 237 (1904)
270. J.J. Thomson, Phil. Mag. **23**, 456 (1912)
271. W. Thomson, An account of Carnot's theory, in *Reflections on the Motive Power of Heat*, ed. by R.H. Thurston (Wiley, NY, 1897)
272. W. Thomson, Proc. R. Soc. **54**, 389 (1893)
273. G.E. Uhlenbeck, S. Goudsmit, Naturwissenschaften **47**, 953 (1925)
274. G.E. Uhlenbeck, S. Goudsmit, Physica (1925)

275. G.E. Uhlenbeck, S. Goudsmit, Nature (Feb. 20, 1926)
276. K. Umenaga, Fukuoka Univ. Ed. III **31**, 49 (1981)
277. C.F. Varley, Proc. R. Soc. **19**, 236 (1871)
278. M. von Laue, Ann. Phys. **18**, 523 (1905)
279. S.R. Weart, M. Phillips (eds.), *History of Physics: Readings from Physics Today* (AIP, New York, 1985)
280. G. Wentzel, Z. Physik **40**, 590 (1926)
281. J.B. West, Physiology (Bethesda), **28**(2), 66–73 (2013). https://doi.org/10.1152/physiol.00053.2012
282. W. Wien, O. Lummer, Ann. der Phys. **56**, 453 (1895)
283. W. Wien, Ann. der Phys. **52**(288), 132 (1894)
284. W. Wien, Ann. der Phys. **58**(294), 662 (1896)
285. W. Wien, Verh. physik. Ges zu Berlin **16**, 165 (1897)
286. W. Wien, Verh. physik. Ges zu Berlin **17**, 110 (1898)
287. A.H. Wilson, *Thermodynamics and Statistical Mechanics* (Cambridge University Press, 1957)
288. E.T. Whittaker, *A Treatise on the Analytical Dynamics of Particles and Rigid Bodies* (Cambridge University Press, London, 1964)
289. E.T. Whittaker, *A History of the Theories of Aether and Electricity from the Age of Descartes to the Close of the Nineteenth Century* (Longmans, Green, and Co., London, 1910) (Reprinted by BiblioLife)
290. A. Whitaker, *Einstein, Bohr and the Quantum Dilemma* (Cambridge University Press, Cambridge, 2006)
291. R. Young, Interview of Paul Peter Ewald by R.A. Young on 1 April 1959, Niels Bohr Library & Archives, American Institute of Physics, College Park, MD, USA, www.aip.org/history-programs/niels-bohr-library/oral-histories/4595
292. P. Zeeman, Reports of the Ordinary Sessions of the Mathematical and Physical Section (Royal Academy of Sciences in Amsterdam) (in Dutch) **5**, 181–184 and 242–248 (1896)
293. P. Zeeman, Phil. Mag., 5th series, **43**(262), 226 (1897). https://babel.hathitrust.org/cgi/pt?id=mdp.39015024088695&view=1up&seq=245
294. E. Zermelo, Annalen der Physik **57**(3), 485 (1896)
295. H. Zinkernagel, Stud. His. Phil. Modern Phys. **53**, 9 (2016)

Index

A
Action, 126
Action integral, 90
Action variable, 91, 150
Adiabatic changes, 11
Albertus University in Königsberg, 87
Alpha-particle, 36
American Physical Society, 135
Ampère, André, 92
Analytical mechanics, 14, 169
Animal electricity, 6
Anion, 8
Anode, 8
Anomalous Zeeman effect, 91
Aristotle, 2, 3
Arnold, Harold D., 132
Arnold Sommerfeld Center for Theoretical Physics, 89
Arosa, Switzerland, 167
AT&T, 132
Atom, 2
Atombau und Spektrallinien, 93
Aurora borealis, 21
Avogadro's number, 71, 72
Azimuthal quantum number, 92

B
Balmer, Johann, 81, 84
Banks, Joseph, 7
Barkla, Charles, 33, 96
Bavarian Academy of Science, 34
Becher, Johann Joachim, 9
Bell Telephone Laboratories, 44, 132, 135
Bergakadamie Clausthal, 88
Bernoulli, Daniel, 4
Berthollet, Claude Louis, 10
Birefringence, 92, 118
magnetically induced, 92
Bohr–Einstein debates, 193
Bohr, Kramers, and Slater (BKS), 112
Bohr magneton, 92, 118
Bohr, Niels, 1, 90, 93, 112
Bohr orbits, 93
Bohr–Sommerfeld theory, 143
Boltzmann equation, 19
Boltzmann, Ludwig, 4, 19, 42, 68
Bormann, Elisabeth, 118
Born, Max, 56, 94, 117, 134, 143, 190
Boscovich, Ruggero Giuseppe, 5, 8, 29
Bothe, Walther, 113
Boyle, Robert, 3
Boyle's gas law, 4
Bragg spectrometer, 96, 110
Bragg, William Henry, 34, 96
Bragg, William Lawrence, 34, 96
Brillouin, Léon , 124
British Association for the Advancement of Science, 13
Brush, Stephen G., 3, 10
Buijs-Ballot, Christophorus, 18
Burschenschaft Germania, 87

C
Caloric, 9–12
Cambridge, 112
Cambridge University, 34
Canonical transormation, 159
Carlisle, Anthony, 7

C. S. Helrich, *The Quantum Theory—Origins and Ideas*, History of Physics,
https://doi.org/10.1007/978-3-030-79268-8

Carnot, Sadi, 12
Carnot's theorem, 12, 15
Cathode, 8
Cathode rays, 25, 79
Cation, 8
Cavendish Laboratory, 24, 33, 35, 82, 132, 143
Cavendish Professor of Physics, 24
Chemical Nomenclature, 10
Clapeyron, Émile, 13
Clausius' inequality, 16
Clausius, Rudolf, 12, 15, 42
  second law, 17
Cloud chamber, 188
College of Wooster, 109
Columbia University, 132
Comptes Rendus, 14
Compton, Arthur H., 109, 113, 114
Compton effect, 111
Compton scattering, 114
Copenhagen, 91, 112
Copenhagen interpretation, 190
Copernican system, 2
Correspondence principle, 112, 176
Count Moltke, 44
Courant, Richard, 140
Crookes' dark space, 23
Crookes, William, 23
Curie, Marie, 33

D
Dancer, John B., 13
Darwin, Charles G., 96
Davisson, Clinton, 131
Davy, Humphry, 7
De Broglie, Louis, 123, 134, 168, 169
De Broglie, Maurice, 123
de Broglie's thesis, 168
de Broglie wavelength, 130
Debye, Peter, 93, 146
Democritus, 2
Dirac, Paul, 140, 145
Displacement operator, 226
Dolezalek electrometer, 107
Dolezalek, Friedrich, 107
Doppler effect, 109
Drude, Paul, 82
Dunoyer, Louis, 117

E
Ehrenfest, Paul, 72
Eidgenössische Technische Hochschule (ETH), 116
Eiffel Tower, 124
Einstein, Albert, 56, 67, 69, 89, 112, 116, 131, 140, 190, 205
Electrode, 8
Electrolyte, 8
Electromagnetic radiation, 43
Electron, 1, 25, 26, 31
  reduced mass in ionized He, 96
Elsasser, Walter, 134
Energy packet, 126
Energy parcel, 126
Ensemble, 27
Ensemble average, 28
Entropy, 16, 17
Epstein, Paul S., 92, 179
Equipartition principle, 70
Euler, Leonhard, 14
Ewald, Paul, 34

F
Fabroni, Giovanni, 7
Faraday, Michael, 8, 21
Faraday's dark space, 22, 23
Fermat's principle, 210
Fermat, Pierre de, 130
Feynman, Richard, 205
First law of thermodynamics, 15, 17
FitzGerald, George F., 1, 23, 25, 30
Fourcroy, Antoine François de, 10
Fourier transform, 35
Fowler, Ralph H., 115
Fowler, Robert H., 163
Franck, James, 103, 134
Frederick of Prussia, 44
French Revolution, 123
Frequency for proper mass, 126
Friedrich, Walter, 34
Friedrich Wilhelm III, 87

G
Galilei, Galileo, 2, 3, 5
Galvani, Luigi, 5
Gay-Lussac, Joseph Louis, 7
Geiger, Hans, 1, 36, 113
Geiger–Marsden experiment, 37
Geissler, Heinrich, 22
General Electric, 132
Geometrical optics description, 176
Gerlach, Walther, 116
Germer, Lester, 132

Gibbs equation, 74
Gibbs, Josiah Willard, 27, 69
Goldstein, Eugen, 23, 26
Gordian knot, 142
Goudsmit, Samuel, 184, 186
Guillotine, 123

H
Hamiltonian, classical, 154
Hamilton–Jacobi equation, 125, 130, 131, 165, 169, 180, 208
Hamilton's principal function, 169
Hamilton, William Rowan, 14, 130, 159, 210
Hartree, Douglas, 134
Harvard Sheldon Fellowship, 112
Hatfield Scholar, 36
Heat, 12
Heilbron, John L., 30
Heisenberg cut, 192
Heisenberg, Werner, 139
Helgoland, 141
Helm, Georg, 42
Helmholtz, Hermann von, 14, 25, 44
Hertz, Gustav, 103
Hettner, Gerhard, 55
Hilbert, David, 87, 140
Hittorf, Johann Wilhelm, 22
Hohlraum, 43
Holton, Gerald, 3, 107
Huygens, Christiaan, 209

I
Ideal gas, 4
Ideal gas law, 11
Indeterminacy principle, 188
Inertial mass, 125
Institute for Theoretical Physics, 117
International Congress of Physicists, 33

J
Jacobi, Carl Gustav, 14, 81, 130, 159
Jammer, Max, 144
Jauncey, George E.M., 110
Jaynes, Edwin T., 59
Jeans, James, 17
Jesuits, 3, 5
Jordan, Pascual, 140, 144
Joule, James Prescott, 13

K
Kármán, Theodore v., 140
Kinetic theory of gases, 42
King's College London, 132
Kirchhoff, Gustav, 43
Klein, Felix, 87
Klein, Martin J., 41, 47
Klein, Oskar, 188
Knipping, Paul, 34
Knott, Cargill, 33
Kragh, Helge, 31, 68
Kramers, Hendrick A., 112
Krönig, August, 17
Kundt, August, 146
Kunsman, Charles, 132
Kurlbaum, Ferdinand, 49, 50, 54

L
Ladenburg, Rudolf, 146
Lagrange, Joseph-Louis, 14, 169
Lagrange undetermined multipliers, 180
Langevin, Paul, 123, 131
Langmuir, Irving, 132
Laplace, Marquis de, 10
Laue, Max, 96
Lavoisier, Antoine, 7, 9, 10
Leiden University, 186
Lenard, Philipp, 25, 79
Leucippus, 2
Light particles, 124
Light waves, 69
Lindemann, Ferdinand, 87
Lommel, Eugen v., 45
Lord Kelvin, 12
Lord Rayleigh, 46
Lorentz, Hendrik Antoon, 90, 186
Lorentz transformation, 127
Loschmidt, Josef, 20
Lucas, Francis F., 134
Ludwig Maximilian University, Munich, 34, 88
Lummer, Otto, 49, 53
Lund University, 84

M
Manchester, England, 82
Manhattan Project, 103
Marsden, Ernest, 1, 36
Maupertuis, Pierre Louis, 14, 130, 169
Maxwell, James Clerk, 4, 24, 32, 69, 190
Mayer, Alfred M., 29
McGill University, 35
Metre Convention of 1875, 44
Michelson, Albert A., 105

Michelson, Wladimir, 45
Miller indices, 35
Millikan, Robert A., 105
Minkowski, Hermann, 127, 128, 139
Mintzer, David, 21, 29
Moltke, Helmuth von, 44
Moore, Walter, 167
Morveau, Guyton de, 10
Moseley, Henry, 96
Munich, 11

N
Nagaoka, Hantaro, 32
National Bureau of Standards, 44
National Physical Laboratory, 44
Natural radiation, 46, 64
Nernst, Walther, 117
Neumann, John von, 153
Newton, Isaac, 4
Nichols, Ernest F., 49
Nicholson, John W., 83, 90
Nicholson, William, 7
Niedersachsen, 144
Normal Zeeman effect, 91
Number mysteries, 94
Numerical harmonies, 94

O
Ostwald, Wilhelm, 42

P
Parmenides, 2
Pascal, Blaise, 3
Paschen, Friedrich, 48, 93, 104, 117
Pauli, Wolfgang, 91, 118, 119, 142, 186
Peierls, Rudolf, 190
Périer, Florin, 3
Phase wave, 128, 130
Phlogiston, 9
Photoelectric effect, 69, 79
Photoluminescence, 78
Photon, 124
Physikalisch-Technische Reichsanstalt, 41
Pickering, Edward Charles, 95
Planck, Max, 41, 42, 67, 109, 190
Planck's second theory, 69
Plücker, Julius, 22
Plum pudding model, 30
Poincaré, Henri, 20
Poisson bracket, 163
Princeton University, 109
Principal function of Hamilton, 14, 199
Principal quantum number, 92
Principia Mathematica, 4
Pringsheim, Ernst, 53
Proceedings of the Copenhagen Academy, 93
Projection quantum number, 92
Proper energy, 125
Proper time, 127
Protyle, 29, 31
Prout, William, 29
Puy de Dôme, 3

Q
Quanta, 125
Quantum relationship, 126
Quantum statistical mechanics, 131
Quantum theory, 134

R
Ramsauer, Carl, 106
Reduced mass, 96
Reeve, Howard, 134
Relativistic dynamics, 125
Relativity, special theory, 69
Ricci, Michelangelo, 3
Richardson, Owen, 131
Röntgen, Wilhelm, 33
Rosenfeld, Léon, 190
Royal Academy of Sciences in Paris, 11
Royal Academy of Sciences in Turin, 14
Royal Institution, London, 7, 25
Royal Society of London, 14
Rubens, Heinrich, 50, 54
Rumford, Count, 11
Rutherford, Ernest, 35, 82, 109
Rydberg constant, 84
Rydberg, Janne, 84

S
Sadler, Charles A., 33, 96
Salviati, 2
Scherrer, Paul, 93
Schrödinger equation, 171
Schrödinger, Erwin, 94, 131, 134, 167, 169
Schuster, Arthur, 25
Schwarz inequality, 222
Scintillation technique, 36
Second law of thermodynamics, 15, 17
Siemens, Werner von, 44
Silvaplana, 145

Simon, Alfred W., 113, 114
Slater, John C., 112
Solvay Conference, 90, 123
Sommerfeld, Arnold, 34, 59, 88, 93, 140, 184, 186
Spatial quantization, 118
Specific heat, 11
Speed of sound, 11
Stahl, Georg Ernst, 9
Stark effect, 92
Stark, Johannes, 80
State vector, 199
Statistical mechanics, 27, 28, 69
Stern, Otto, 116
Stokes' law, 78
Stoney, George J., 25
Strutt, John W., 46

T
Thénard, Louis Jacques, 7
Theorem of phase harmony, 127
Thermochemistry, 10
thermodynamics, second law, 42
Thermodynamic temperature, 12
Thermodynamic work, 13
Thiesen, Max, 54, 61
Thompson, Benjamin, 11
Thomson, Joseph John, 1, 24, 25, 30, 43, 79, 82, 109, 132
Thomson, William, 12, 13, 17, 25, 30
Tokyo University, 33
Torricelli, Evangelista, 3
Trace of a matrix, 152
Tripos examination, 35

U
Uhlenbeck, George, 184, 186
Ultraviolet catastrophe, 72
Unitary transformation, 225
University of Basel, 84
University of Bologna, 5
University of Breslau, 116
University of California at Berkeley, 132
University of Chicago, 103, 189
University of Copenhagen, 82
University of Frankfort, 117
University of Göttingen, 103
University of Leeds, 96
University of Manchester, 35
University of Minnesota, 109
University of Pavia, 6
University of Prague, 116
University of Tübingen, 117, 145
University of Zurich, 116

V
Vacuum, 2
Variational calculus, 171
Varley, Cromwell, 23
Volta, Alessandro, 6
Voltaic pile, 7
Von Laue, Max, 1, 34

W
Washington University, 110
Watson, William, 21
Wave function, 171
Weiss, Pierre, 118
Wentzel, Gregor, 189
Western Electric, 132
Westinghouse, 109
Weyl, Hermann, 94, 171
Whitaker, Andrew, 190, 192
Wien, Wilhelm, 26
Wilson, Charles Thomson Rees, 115
World War 1, 123
Wrangler, 35

X
X-ray diffraction, 35
X-rays, 124, 125

Y
Yale College, 27

Z
Zeeman effect, 90, 93
Zeeman, Pieter, 90, 92
Zermelo, Ernst, 20, 42
Zinkernagel, Henrik, 192

9783030792671